Springer **M**onographs in **M**athematics

More information about this series at http://www.springer.com/series/3733

Timothy D. Andersen • Chjan C. Lim

Introduction to Vortex Filaments in Equilibrium

Timothy D. Andersen
Georgia Tech Research Institute
Georgia Institute of Technology
Atlanta, GA, USA

Chjan C. Lim
Department of Mathematical Sciences
Rensselaer Polytechnic Institute
Troy, NY, USA

ISSN 1439-7382 ISSN 2196-9922 (electronic)
ISBN 978-1-4939-1937-6 ISBN 978-1-4939-1938-3 (eBook)
DOI 10.1007/978-1-4939-1938-3
Springer New York Heidelberg Dordrecht London

Library of Congress Control Number: 2014951028

Mathematics Subject Classification: 76B47, 82B99, 82D10

© Springer Science+Business Media New York 2014
This work is subject to copyright. All rights are reserved by the Publisher, whether the whole or part of the material is concerned, specifically the rights of translation, reprinting, reuse of illustrations, recitation, broadcasting, reproduction on microfilms or in any other physical way, and transmission or information storage and retrieval, electronic adaptation, computer software, or by similar or dissimilar methodology now known or hereafter developed. Exempted from this legal reservation are brief excerpts in connection with reviews or scholarly analysis or material supplied specifically for the purpose of being entered and executed on a computer system, for exclusive use by the purchaser of the work. Duplication of this publication or parts thereof is permitted only under the provisions of the Copyright Law of the Publisher's location, in its current version, and permission for use must always be obtained from Springer. Permissions for use may be obtained through RightsLink at the Copyright Clearance Center. Violations are liable to prosecution under the respective Copyright Law.
The use of general descriptive names, registered names, trademarks, service marks, etc. in this publication does not imply, even in the absence of a specific statement, that such names are exempt from the relevant protective laws and regulations and therefore free for general use.
While the advice and information in this book are believed to be true and accurate at the date of publication, neither the authors nor the editors nor the publisher can accept any legal responsibility for any errors or omissions that may be made. The publisher makes no warranty, express or implied, with respect to the material contained herein.

Printed on acid-free paper

Springer is part of Springer Science+Business Media (www.springer.com)

Preface

This book is an introduction to vortex filaments statistics in equilibrium. It is intended to be a survey of applications, methods, mathematical models, and computational methods for vortex filaments in ensembles wherever they appear focusing on hydrodynamics, plasmas (magnetohydrodynamics), and quantized vortices in superfluids and superconductors. The goal is to bring these distinct cases under the single umbrella of vortex statistics. We pay considerable attention to applications in this book, more than is typical in many mathematical treatments. These applications serve as motivation for the models we examine.

This book is intended to be for a graduate seminar or survey course. It is also intended to be a broad reference on a variety of topics in vortex filaments. Typically, in these courses one or two large projects are ideal for assignments. The book emphasizes understanding over computation, but, occasionally, we will ask questions or suggest problems to solve. The student should be familiar with calculus, vector calculus, differential equations, and partial differential equations, as well as know at least some programming. We do not assume prior knowledge of statistical mechanics.

This book differs from Lim and Nebus [90] in that it focuses on vortex filaments in several application areas rather than only hydrodynamics, including plasmas and quantum filaments. Also, while the previous book was directed at undergraduates and beginning graduate students, this book is at a more advanced level, focusing on graduate students and researchers.

Atlanta, GA, USA Timothy D. Andersen
Troy, NY, USA Chjan C. Lim

Notation

Ordinary vectors are usually given in bold, $\mathbf{v}$, while unit vectors have a caret $\hat{}$. Often we will use the unbolded letter to stand for the magnitude, e.g. $B = |\mathbf{B}|$. Integrals use the typical physicist's notation $\int d^d x f(\mathbf{x})$ where $\mathbf{x}$ is the d-vector variable of integration, d is the dimension, and $f(\mathbf{x})$ is the integrand. Line integrals are given by $\oint \mathbf{F} \cdot d\mathbf{l}$. Derivatives of the form $\partial f/\partial x$, f_x, $\partial_x f$, and f' are all equivalent. For simplicity we will also sometimes write $f(x)$ as f with the understanding that these are equivalent. If $\mathbf{F}$ is a vector valued function, then F_x is the x component rather than the derivative. We will use divergence, $\nabla \cdot \mathbf{F} = \partial_x F_x + \partial_y F_y + \partial_z F_z$, often.

Acknowledgements

We would like to acknowledge the National Oceanic and Atmospheric Administration (NOAA), the National Aeronautics and Space Administration (NASA), the National Science Foundation (NSF), and others for making available essential resources, data, and technologies without which this book would not have been possible. The second author would like to thank the generous support of Department of Energy (DE-FG02-04ER25616) and the US Army Research Office (ARO W911NF-05-1-0001 and W911NF-09-1-0254) over the past 10 years. We would also like to thank our families for their support.

Contents

1. **Introduction** 1
2. **Vortex Filaments and Where to Find Them** 9
   - 2.1 Hydrodynamics 9
     - 2.1.1 Planetary Atmospheres 10
     - 2.1.2 Aerodynamic Problems 13
     - 2.1.3 Bathtub Vortex 15
     - 2.1.4 Deep Ocean Convection 15
   - 2.2 Quantized Vortices 16
     - 2.2.1 Bose–Einstein Condensates 17
     - 2.2.2 Superconductors 18
   - 2.3 Plasmas 20
     - 2.3.1 Magnetohydrodynamics 21
     - 2.3.2 Atmospheric Plasma 23
     - 2.3.3 Solar Dynamics 23
     - 2.3.4 Astrophysical Plasmas 23
   - 2.4 Concluding Remarks 24
3. **Statistical Mechanics** 25
   - 3.1 History 26
   - 3.2 Ensembles 27
   - 3.3 Assumptions 31
   - 3.4 Methods 32
     - 3.4.1 The Variational Approach 32
     - 3.4.2 The Integral Approach 37
   - 3.5 Statistical and Fluid Mechanics 40
4. **Parallel Filaments** 43
   - 4.1 The Point Vortex Gas 44
   - 4.2 Negative Temperature States 47

ix

	4.3	The Guiding Center Model	51
	4.4	Continuous Vorticity	54

5 Curved Filaments ... 57
- 5.1 Motion of a Vortex Filament ... 57
  - 5.1.1 The Curvilinear Formulation ... 59
  - 5.1.2 A Small Example of Matched Expansion ... 62
  - 5.1.3 Matched Equations ... 63
- 5.2 Nearly Parallel Vortex Filaments ... 65
- 5.3 Filament Crossing and Reconnection ... 67
- 5.4 Concluding Remarks ... 69

6 Quantum Fluids ... 71
- 6.1 Bose–Einstein Condensates ... 71
  - 6.1.1 Modeling Quantized Vortex Lines in BECs ... 72
- 6.2 Superconductors ... 75

7 Plasmas ... 81
- 7.1 Magnetohydrodynamics ... 81
- 7.2 Confined 2D Plasmas ... 83
- 7.3 Quasi-2D Electron Columns ... 83
  - 7.3.1 A Mean-Field Approach ... 86
  - 7.3.2 A Variational Approach ... 90
- 7.4 Interpretation of Negative Specific Heat ... 96

8 Computational Methods ... 99
- 8.1 Numerical PDEs ... 99
- 8.2 Canonical Ensemble ... 101
  - 8.2.1 Monte Carlo ... 101
  - 8.2.2 Metropolis ... 102
  - 8.2.3 Path Integral Methods ... 103
- 8.3 Microcanonical Ensemble ... 105
  - 8.3.1 Demon Algorithm ... 106
  - 8.3.2 Hamiltonian Flow ... 108
- 8.4 Numerical Simulations in Vorticity ... 109
- 8.5 Concluding Remarks ... 109

9 Quasi-2D Monte Carlo in Deep Ocean Convection ... 111
- 9.1 The Nearly Parallel Vortex Filament Model's Entropy-Driven Shift ... 113
  - 9.1.1 Background ... 113
  - 9.1.2 Hypothesis ... 114
  - 9.1.3 Mathematical Model ... 114
- 9.2 A Mean-Field Approach ... 115
- 9.3 Solving for the Square Radius ... 117
  - 9.3.1 Harmonic Oscillator Approach ... 117
  - 9.3.2 Spherical Method Approach ... 120

	9.4	Monte Carlo Comparison 123
	9.5	Related Work 126
	9.6	Conclusion 126

10 Conclusion 129

References 131

Index 137

Chapter 1
Introduction

Vortex filaments and vortex dynamics have been an important subfield of fluid mechanics since Helmholtz's 1858 paper (translated to English by P. G. Tait) "On the Integrals of the Hydrodynamical equations, which Express Vortex Motion" and the work of William Thomson (Lord Kelvin) [143] and has since become a field of study in its own right. A vortex filament is a thin curve or line of vorticity embedded within a fluid and arises mainly because of rotational motion such as convection currents. In order for a vortex to be a filament it needs to obey the asymptotic relation: $\delta \ll R$ where δ is the core size and R is the curvature. A common further constraint can be placed, $R \ll 1$, in which case the filament is nearly straight as well.

In this book we will cover vorticity in several domains, including magnetohydrodynamics and quantum fluids, but here we will give a derivation for hydrodynamics. Vortex filaments come from the Navier–Stokes equations. Let $\omega = \nabla \times \mathbf{u}$ be the vorticity where $\mathbf{u}(\mathbf{x},t)$ is the fluid velocity localized in some domain D. The vorticity is given by

$$\frac{\partial \omega}{\partial t} = \nabla \times (\mathbf{u} \times \omega) + \nu \nabla^2 \omega,$$

where ν is viscosity. The first term on the right-hand side is the convection while the second term is the dissipation. We are primarily interested in fluid evolution where convection dominates (if dissipation dominates, as it does inside animal cells, then vorticity is not a concern). If fluid behavior is dominated by large-scale nonlinear interactions and dissipation is negligible, we can let $\nu \approx 0$, and use the Euler equations for ideal fluids,

$$\frac{\partial \omega}{\partial t} = \nabla \times (\mathbf{u} \times \omega).$$

Vorticity is said to be localized if its magnitude decreases super-exponentially outside a bounded region, i.e., $\omega(\mathbf{x},t)$ is localized in D at time t if there exists an $d > 0$ such that

$$\lim_{|\mathbf{x}| \to \infty} \left| e^{\mathbf{x}/d} \cdot \omega(\mathbf{x},t) \right| = 0.$$

Also, if vorticity is localized at a time t_0 it is localized at any instant $t > t_0$. Let vorticity be localized.

Given that $\omega = \nabla \times \mathbf{u}$ and the fluid is divergence free, $\nabla \cdot \mathbf{u} = 0$, we can define a vector potential $\psi(\mathbf{x},t)$ such that $\mathbf{u} = \nabla \times \psi$ and $\nabla \cdot \psi = 0$, where ψ limits to a constant at infinity. Since,

$$\omega = \nabla \times (\nabla \times \psi) = \nabla(\nabla \cdot \psi) - \nabla^2 \psi,$$

the vorticity is given by Poisson's equation (in vector form),

$$-\omega = \nabla^2 \psi.$$

Poisson's equation has a solution,

$$\psi(\mathbf{x},t) = \frac{1}{4\pi} \int_\Omega d^3 x' \frac{\omega(\mathbf{x}',t)}{|\mathbf{x}-\mathbf{x}'|},$$

where Ω is a closed domain over which the vorticity is defined. Taking the curl, we get the fluid velocity,

$$\mathbf{u}(\mathbf{x},t) = \nabla \times \psi = \frac{1}{4\pi} \int_\Omega d^3 x' \frac{\omega(\mathbf{x}',t) \times (\mathbf{x}-\mathbf{x}')}{|\mathbf{x}-\mathbf{x}'|^3}. \tag{1.1}$$

Equation (1.1) is the Biot–Savart law of vortex dynamics in hydrodynamics [131] and can be generalized to magnetohydrodynamics (where the vorticity is a combination of vorticity and magnetic field discussed in Chap. 7). This law is the starting point for all vortex–vortex interactions and is critical to all studies of vortices in equilibrium.

Helmholtz gives three laws of vortex motion for an ideal, barotropic fluid being acted on by conservative external forces [131],

1. Fluid particles that are irrotational remain irrotational.
2. Fluid particles on a vortex line at any time remain on that vortex line for all past and future times.
3. The strength of a vortex tube is constant with respect to the fluid motion.

The first law is analogous to the conservation of particle number because it says that vortices cannot be created out of nothing. In other words, just as particles cannot appear in the universe out of nothing, because irrotational particles of the fluid remain irrotational, a vortex cannot simply appear in a fluid spontaneously. The second law means that vortex lines move with the fluid, i.e., if a particle moves, the vortex line moves with it so the particle always stays on the line. This law is analogous to the law of inertia or conservation of momentum. We know from general relativity that particles move with space and time, i.e., space and time act, in some respects, like a fluid medium. The conservation of momentum says that particles move with this medium. In the same way, a vortex line cannot change direction or move in a direction contrary to its medium. The third law is analogous to the conservation of charge/mass since vortex strength is similar to charge/mass in that it governs the strength of vortex interactions.

1 Introduction

Once strong equivalences between vortex lines and ordinary particles were created, it made sense to study vortex gases, liquids, and solids as phenomena in their own right. Statistical equilibrium ensembles of vortex lines or points, however, were not considered until the 1940s when Lars Onsager calculated negative temperature states for fixed energy ensembles of vortex points in a circular container [117]. While, prior to his discovery, negative temperature states were believed to be unphysical, Onsager proved that above a certain maximal energy the temperature becomes negative and vortices of like sign, which at lower energies repelled one another, now attracted one another. This discovery encouraged continued study, and, in the 1970s, Edwards and Taylor [46] and Joyce and Montgomery [70] showed that Onsager's results also applied to magnetically confined plasmas where columns of electrons interacted in the same way as vortex lines, including having negative temperature states. Because these plasmas provide a potential route to hot magnetic nuclear fusion, the study of vortex filaments has since intensified.

Statistical mechanics forms the basis for the study of large numbers of vortices. Although macroscopic, systems or ensembles of vortex filaments have state variables such as temperature, entropy, pressure, specific heat, etc. which are the result of their combined interaction. These state variables, although statistical, become, in the limit as the number of vortices becomes large, exactly related to one another through equations of state (e.g., the ideal gas law for non-interacting particles, $PV = nRT$, also applies to non-interacting vortex lines, albeit in two dimensions).

In this book we will deal exclusively with equilibrium states, although we will occasionally touch on what happens when equilibrium states are not stable. What counts as a stable state and how long is enough time depend on the system and are often open problems. In the case of gases in a box, the quintessential subject of statistical mechanics, stability can be indefinite provided the box is either completely insulated from the outside world or in a stable heat bath. Gas molecules are extremely stable, of course, particularly in isolation. With more exotic types of particles that tend to decay or change over time, outer stability is no guarantee of inner stability. In many systems, the so-called stable states can be very short before outside influences bring the system to a different state. Vortices, in particular, tend to dissipate. They can, however, form configurations within a short time period that can be described by equilibrium statistics.

One of the major early discoveries in the study of interacting vortices is that they behave a lot like charged particles. When vortex lines are assumed to be perfectly straight, parallel, and infinitesimally thin, equations of state may be derived from Poisson's two-dimensional equation,

$$\nabla^2 \phi = -4\pi\rho, \tag{1.2}$$

relating the interaction potential to the density of vorticity, ρ, and the energy functional,

$$E = \int d^2x \int d^2x' \rho(\mathbf{x})\rho(\mathbf{x}') \log|\mathbf{x}-\mathbf{x}'|. \tag{1.3}$$

These are, of course, exactly the equations that govern interacting charged particles, although the kinetic energy is conspicuously absent from the above functional. In the idealized case of perfectly straight filaments, the kinetic energy can be shown to be infinite. In reality, vortices are never perfectly straight, but, when they are very nearly straight, their kinetic energy is large and virtually constant. Since only changes in energy from state to state matter in statistics, the kinetic energy can be neglected. We will address the case where vortices are not straight enough in Chap. 5.

Although they have a simple interaction, two-dimensional vortex filaments and points are interesting because they are complex enough, even without complex boundaries, to be unsolved and perhaps unsolvable in statistical equilibrium in the case of N vortices. For instance, the partition function at an inverse temperature, β, of an ensemble of like-signed vortices in a harmonic trap is given by,

$$Z = \int dz_1 \cdots dz_N \exp\left[-\beta\left(\sum_{i=1}^{N} |z_i|^2 - \sum_{i<j} \log|z_i - z_j|\right)\right], \quad (1.4)$$

where $z_j = x_j + iy_j$ is a complex number representing the position of vortex j. While the one-dimensional N vortex problem has been solved for nearly four decades [52], despite the efforts of numerous mathematicians and physicists, this two-dimensional problem remains solved for only a few values of $N = 1, 2, 3, 4$ and one value of β, $\beta = 2$ [55]. Without having an exact solution that is easy to calculate, researchers must rely on various mathematical "tricks," approximations, or else simulate the system directly to understand its behavior. Additionally, and perhaps more tantalizing mathematically, this partition function also describes the statistics of eigenvalues of random matrices where β describes from which distribution the elements are drawn. Since this book is not about random matrices, we will not go into this, but it does suggest a deep connection between vortices, charged particles, and random matrices. For a discussion of random matrices, see Mehta's book [104].

In this book, we make a fundamental assumption in order to apply statistical mechanics to vorticity: that vortices are particles. In statistical equilibrium studies of any type of particle, stars, molecules, or vortices, the internal structures of those particles are necessarily ignored, and they are assumed to be localized and self-contained. In this book, the assumption of localized, self-contained particles will be assumed *a priori* unless otherwise specified. This assumption is not always justified. A subatomic particle, for example, is not self-contained at the quantum level, and one of the great open problems of modern physics is how to justify classical statistical mechanics given that the superposition of quantum states is possible (quantum decoherence is one proposed solution). In other words, because particles are not localized but can appear to exist in more than one place at the same time, the assumption that any given particle does not have wave-like interference with itself is not justified in all cases. A good example of a situation where quantum mechanics would interfere with classical statistical mechanics is in cases of nonlocality such as the Aharonov–Bohm effect where electrons are affected by the presence of distant or insulated electromagnetic fields. Another would be in the case of rapidly decaying particles where equilibrium can only be attained over short time scales.

This problem also applies not only to tiny particles but also to the statistics of very large bodies. In the treatment of catastrophic gravo-thermal collapse in globular clusters, the internal structures of stars are ignored as well even though stars continually release charged particles and often have clouds of material surrounding them up to great distances, they too are assumed to be self-contained, again an assumption that is not justified if the stars are too close together or undergoing significant change (such as collapsing or expanding). Vortices are no different. Like subatomic particles, they may have wave-like interference with themselves. They can also disperse or merge.

All statistical mechanical ensembles have a range of validity, and it is important that such phenomena as particle merger, non-localized effects such as self-interference, and particle dispersion be minimal for the time-scale and parameters chosen. When one or more of these effects ceases to be irrelevant, the statistical model may fail as in the case of low-temperature models of liquid helium or alkali atoms where quantum mechanical effects (such as self-interference) become significant near absolute zero as the atoms fall into a single quantum state. In these cases, a new statistical ensemble may be formulated taking the new effect into account.

When vortex filaments behave like particles is an important question, and the answer is that it depends on the medium. In ordinary fluids, vortex filaments have the least longevity and are most likely to merge and break because of viscosity. In this case, time scales must be very short or, alternatively, vortex filaments must be treated as idealized particles of angular momentum. In this latter case, the filaments themselves are not real but represent infinitesimal amounts of the angular momentum of a rotating fluid. This interpretation is a way to discretize a larger vortex into many small filaments so that the larger vortex can be modeled effectively. With this interpretation, filaments can be presumed to persist as long as the larger vortex does, which can be a substantial amount of time. In superfluids and superconductors, which lack viscosity, vorticity and magnetic fields, respectively, are explicitly quantized and Abrikosov lattices distinctly visible. Here, vortex filaments persist as discrete entities for long periods of time. In dilute Bose–Einstein Condensates of alkali atoms which become superfluidic below about 50 nK [1] this is even more true.

That vortex filaments appear quantized in superfluids, even though superfluids are, by definition, irrotational, indicates how fundamental vortex filaments are and why treating them as particles has become so common. Indeed, in this case, they are a direct consequence of the quantization of spin which is fundamental to the structure of matter in the universe. In non-quantum fluids, plasmas, etc. vortex filaments appear as particles on short time-scales. For example, a rake drawn through a fluid (a common apparatus in tabletop turbulence experiments) creates a set of vortex filaments, which may then merge and/or dissipate through friction against the boundary walls. These filaments are unforced and last for only a short time. The time-scale, however, is still large enough for the vortices to reach an equilibrium state provided that the inverse energy cascade and the boundary dissipation time-scales are sufficiently long. This is often the case. Without energy dissipation of any kind, indeed, the vortices would continue forever, so the time-scale for which the equilibrium statistical model is applicable is a function of the details of the experiment, e.g., the type of fluid, temperature, the smoothness of the boundary walls, etc. On an appropriate time scale, vortices in ordinary fluids can be treated much like their quantized counterparts, albeit they are described by the Euler or Navier–Stokes vorticity equations rather than, for example, the Gross–Pitaevskii equation for quantum fluids [121]. In other words, whether vortex filaments meet the conditions of being particles and, more importantly, particles in equilibrium depends fundamentally on the time-scale of observation and the speed at which they dissipate. Other factors that may influence the time-scale include the vortex circulation (speed of rotation), strength of any trapping field, counterflows, and whether filaments of opposite sign are present (which can lead to collisions) all of which can increase or decrease the stability of the filaments.

One of the novel chief focuses of this book is the study of quasi-2D vortex filament statistics (Fig. 1.1). Often when vortices are treated as particles, they are treated as points. Recently, limitations to the most common form of vortex filament statistics, 2D point vortices such as Onsager introduced, have become increasingly clear. Because these ensembles have no kinetic energy, they experience stable states that no true system would ever experience. For example, based on statistical

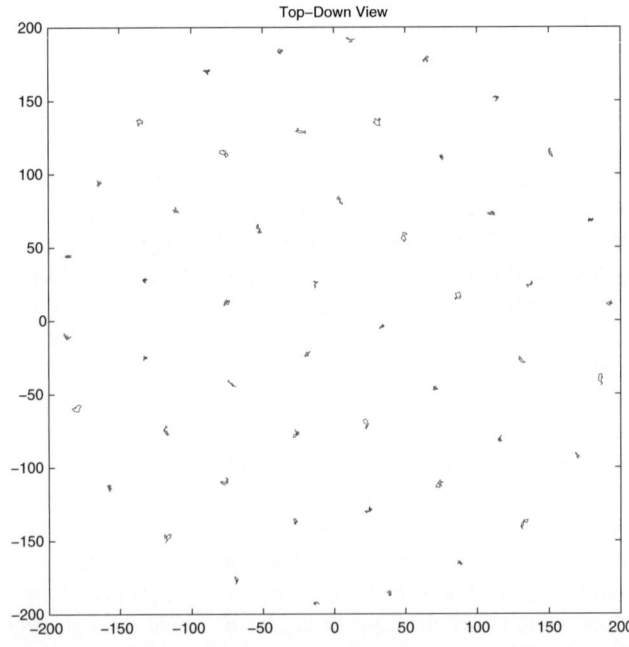

Fig. 1.1: Shown here in *top-down* projection, nearly parallel vortex filaments, at low density and high strength of interaction, are well ordered into a 2-D triangular lattice known as the Abrikosov lattice from type-II superconductors [2]. This figure (taken from a single Monte Carlo sample) shows how the quasi-2D model is essentially a 2-D model for these parameters

analysis, an ensemble of identical point vortices always reaches a preferred radius for a fixed energy. Because energy decreases the farther like-signed vortices are from one another (because they repel one another), that radius cannot be exceeded as long as kinetic energy is ignored. Of course, no true system of vortices in an ordinary fluid stays at a fixed size. It disperses, even in the absence of boundary dissipation or other heat flux, because the third dimension is an energy reservoir where potential can convert into kinetic energy. Therefore, an ensemble of identical vortex filaments with kinetic energy, even at fixed energy, can continue to expand far beyond the fixed size of the purely 2D ensemble. This is what we would expect, of course. After all, if we take away the kinetic energy of a falling object but leave it its gravitational potential, at fixed energy, it is unable to fall because conversion of potential to kinetic energy is impossible. Thus, removing the kinetic energy of the object leaves it hanging in mid-air. Clearly, 2D point vortex models have their usefulness, but they neglect important considerations as well.

Indeed, while the 2D ensemble exhibits negative temperature states, it is incapable of exhibiting negative specific heat states, which are so common to gravitational systems, because it lacks a kinetic energy. The specific heat of the 2D ensemble at fixed volume circumscribed by a circle of radius R is, in fact, always positive. On the other hand, quasi-2D filaments have negative specific heat when held in a magnetic trapping potential. Negative specific heat in vortex systems is particularly important both for studying turbulence in ordinary fluids and magnetic plasma confinement since it implies that these ensembles are exactly opposite to what is expected of any statistical ensemble.

Negative specific heat, in fact, is a much more bizarre phenomenon than negative temperature states (which flips the sign of the Coulomb force from repulsive to attractive for like-signed filaments) since it implies that adding energy to a system cools it. No one has ever put a pan of water over a fire and seen it freeze, so this is beyond ordinary experience. In gravitational systems negative specific heat is directly related to the core collapse of globular clusters where a decrease of the gravitational potential (core collapse) causes an increase in temperature (because stars move faster the closer they are to the center of mass). If such a phenomenon also occurs in vortex systems and the conditions for it could be discovered more precisely, core collapse of plasma ions could cause them to fuse, leading to a sustainable fusion reaction [73]. (A more detailed discussion of negative specific heat appears in Chap. 7.)

This book is organized as follows: Chap. 2 is devoted to the appearance of vortex filaments and ensembles of many vortices in the natural world and includes discussions of ordinary fluids, quantum fluids, type-II superconductors, magnetically confined plasmas, and deep ocean/atmospheric convection. Chapter 3 provides background on statistical mechanics and particularly its application to vortex filaments. Those well versed in statistical mechanics and mathematical approaches such as how to derive equations of state from a statistical ensemble, variational versus integral methods, mean-field theory and the like may want to skip this chapter. Those who want to skip straight to the mathematics should begin with Chap. 4, which discusses parallel vortex filaments, including Onsager's contributions, their relationship to plasmas, and continuum theories. Chapter 5 discusses curved vortices, including nearly parallel vortex filaments and how they are derived asymptotically. Chapter 8 discusses computational approaches to vortex filaments. Chapters 6 and 7 discuss the two main non-hydrodynamical applications of vortex filaments, quantum fluids, and plasmas. Chapter 9 provides a lengthy discussion of a particular case of an ensemble of nearly parallel vortex filaments, combining both analytical mean-field methods and Monte Carlo, to model a problem in deep ocean convection.

Chapter 2
Vortex Filaments and Where to Find Them

In this chapter, we will look at the different places that one might find vortex filaments. In later chapters, we will go more deeply into the analysis of these filaments.

Unlike molecules, isolated vortex filaments have been observed for millennia. Whenever the Reynolds number, a measure of the ratio of inertial to viscous force, of a fluid exceeds a certain value (depending on the type of flow) no steady flow is stable [117], and in these unsteady flows vortices form. Whirlpools, convection currents, tornadoes, hurricanes, and swirls in a draining tub are all examples of vortex filaments or ensembles of filaments. Vortex-like filaments also occur in electron plasmas, magnetohydrodynamic fluids, the solar atmosphere, type-II (high temperature) superconductors, superfluids, and Bose–Einstein condensates. In this chapter, we discuss a variety of examples of vortex filaments in nature to motivate the scientific relevance of the mathematical results in the rest of the book.

2.1 Hydrodynamics

Hydrodynamics is a large subfield of physics and applied mathematics that is concerned with the description of ordinary, neutral fluids, usually water although theoretical models can apply to nearly any fluid, provided it is non-relativistic and uncharged, including air, oil, and artificial fluids with customized viscosities and other properties. These fluids are called ordinary or "Navier–Stokes" fluids because they are described by the Navier–Stokes equations, [82],

$$\frac{\partial \mathbf{u}}{\partial t} + \mathbf{u} \cdot \nabla \mathbf{u} = -\frac{\nabla P}{\rho} + v \nabla^2 \mathbf{u}, \qquad (2.1)$$

where $\mathbf{u}(\mathbf{x},t)$ is the velocity field, $\rho(\mathbf{x},t)$ is the density, $P(\mathbf{x},t)$ is the pressure field, and v is the viscosity constant. As mentioned in the introduction, the Navier–Stokes is the gold standard of hydrodynamical models but by no means the most useful.

© Springer Science+Business Media New York 2014
T.D. Andersen, C.C. Lim, *Introduction to Vortex Filaments in Equilibrium*,
Springer Monographs in Mathematics, DOI 10.1007/978-1-4939-1938-3_2

Many simpler and more applicable models can be derived via assumptions (*Ansatz*) about the nature of the fields the equation contains and/or via asymptotic analysis.

The vorticity form of the Navier–Stokes is straightforward. Vorticity is the amount of rotation or *curl* of the velocity field. Let $\omega = \nabla \times \mathbf{u}$ be the vorticity field, then the vorticity equation is

$$\frac{D\omega}{Dt} = \omega \cdot \nabla \mathbf{u} + \nu \nabla^2 \omega \tag{2.2}$$

where $D/Dt = \partial/\partial t + \mathbf{u} \cdot \nabla$ is the total derivative. From this equation one may derive the behavior of all vortex motion in a Navier–Stokes fluid. When $\nu = 0$ the fluid is said to be an ideal, inviscid, frictionless, or *Euler* fluid (all these names are equivalent). A vortex is a special feature of vorticity where vorticity is confined to a compact area. Vortices have circulations defined as the line integral of the velocity around a closed curve $\Gamma = \oint_C \mathbf{u} \cdot d\mathbf{l}$. Provided the closed curve completely encircles a vortex filament, the circulation will be independent of the curve.

Whether a fluid can be modeled as an Euler fluid or a Navier–Stokes fluid depends on its Reynolds number, $\mathrm{Re} = vL\rho/\nu$. The variables ν and ρ have already been defined as the viscosity and density, v is the average speed, and L is the length scale. Large, fast objects like airplanes operate at high Reynolds numbers while small slow objects, like insects at slow speeds and microorganisms, operate at small Reynolds numbers. This is why, for example, small creatures move using flagellum, which operate like corkscrews, while large objects are able to move by reciprocal flapping (which does not work at low Reynolds numbers). Reciprocal flapping works because it creates a string of vortices known as a von Karman Street (named for the famous aeronautical scientist Theodore von Karman) which work to propel it forward. At low Reynolds numbers these vortices are damped too quickly to form and no forward motion occurs. The reverse situation to flapping can also occur when motion of the fluid around an object causes the Von Karman Street to be shed on the other side. This often occurs when wind passes around an island (Fig. 2.1).

Although Von Karman vortices are examples of generic vortices, they are not filaments. Waterspouts, essentially tornadoes over water, however, are thin enough to be considered filaments Fig. 2.2.

2.1.1 Planetary Atmospheres

Planetary atmospheres are a good place to observe vortex filaments and other vortical structures [140] and [89] have applied statistical vortex methods to study several aspects including (1) the super-rotation of the atmospheres of Venus and Titan, (2) anticyclonic structures in gas giants such as Jupiter's Great Red Spot, and (3) polar vortices which occur on Earth, Venus, and Saturn.

Atmospheric super-rotation is an unusual phenomenon where an atmosphere on a rotating planet rotates faster than the planet, and, while complex Navier–Stokes models can be applied, this can be explained much more simply by vortex methods.

2.1 Hydrodynamics

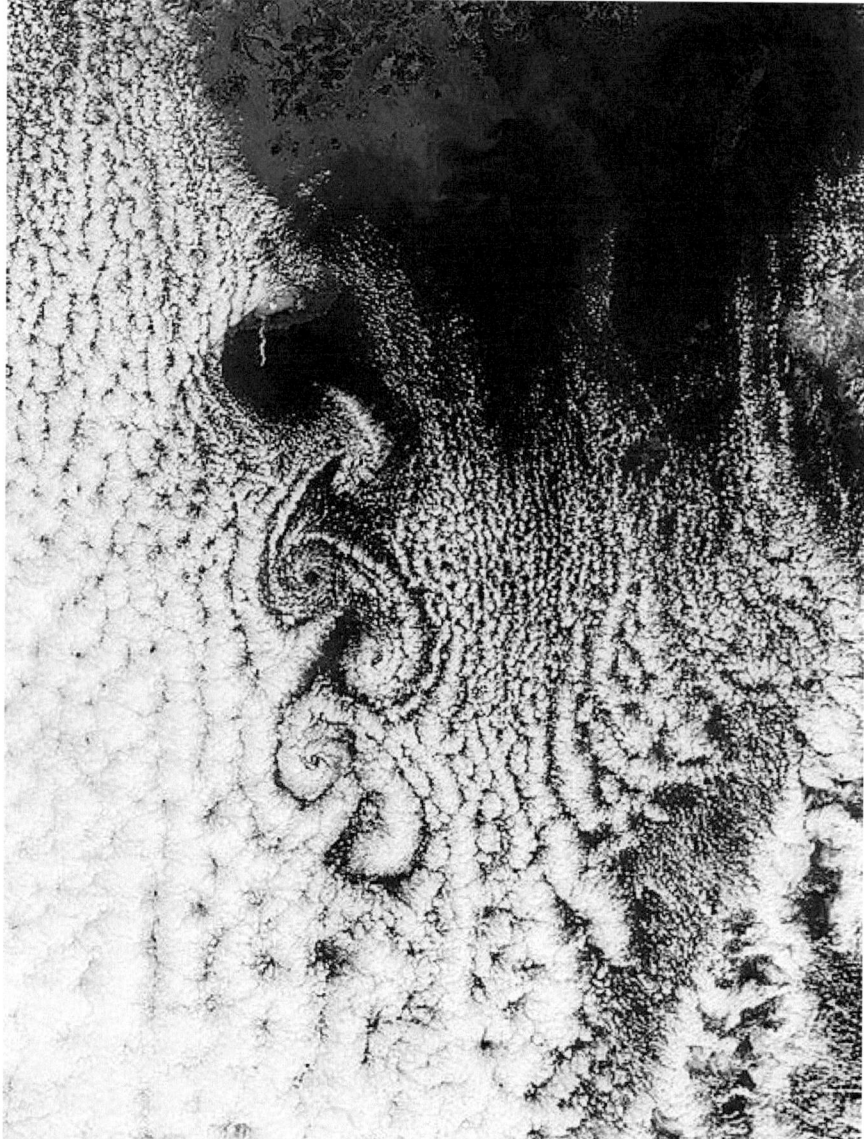

Fig. 2.1: These cloud vortices are an example of a von Karman street off Cheju Do, South Korea. They are not filaments however. *Courtesy of NASA*

Several spacecraft have been sent to Venus and studied its atmosphere including Mariner 10 in 1974, the Pioneer Venus Orbiter in 1978, and Venus Express in 2006. With each mission, the complex structure of the planet's thick atmosphere is further revealed. One of the earliest observations showed that while the planet rotates

Fig. 2.2: This waterspout photographed off the Florida Keys in 1969 is an example of a hydrodynamic vortex filament. *Courtesy of NOAA. Photo by Joseph Golden*

slowly, at once every 243 Earth days, the atmosphere rotates every 4 Earth days. Titan, a moon of Saturn and quite far from the Sun, meanwhile, has been visited several times over the past three decades. Spacecraft include Pioneer 10, which showed Titan to be extremely cold, Voyager 1, which showed it to have an atmosphere 300–350 km thick, and Cassini which showed super-rotation of its atmosphere. The Huygens probe that Cassini dropped into the Titan atmosphere in 2004 also revealed that the atmosphere has several layers which appear to rotate in different directions, west-to-east in the upper layer, then east-to-west, and west-to-east again. Statistical vortex methods applied to a barotropic (single pressure) model, including an exact, closed-form solution, showed phase transitions such that slowly rotating planets can only exhibit super-rotation in the direction of the planet's rotation while at a critical rotation rate the planet can have anti-rotation as well. In this model, the planet's atmosphere is divided up into barotropic vortex cells, which in this case are 2-D but still represent straight filament structures, and these cells are allowed to interact with their neighbors. From this interaction, super-rotation and the phase transition are emergent phenomena. Multi-layered 2D models, meanwhile, can explain why super-rotation and anti-rotation occur at different layers because of changing density, temperature, and viscosity. (Indeed, the standard atmospheric model for the Earth, used by aeronautical engineers and rocket scientists, is a multi-layer model.)

Anticyclonic vortex structures have been observed on several planets including the most famous, Jupiter's Great Red Spot, but also Saturn's Great White Spot,

2.1 Hydrodynamics

and Neptune's Great Dark Spot, and vortex methods yield important clues to their origins. In each of these cases, the spots are not hurricanes with well-defined eye and eye-wall but more like deformations in atmospheres of these planets, defects resulting from complex vortex interactions. The Jovian planets are rapidly rotating planets where the active fluid layer has a dynamic height. A model, known as the "Shallow Water model" because it is designed to represent thin layers of fluid sitting over a harder layer, revealed similar vortex structures. In this case, the rapid spin of the planet and the strong horizontal nature of the flows helped explain, in terms of fluid angular moment and varying surface height, these large-scale structures.

Polar vortices are the another large atmospheric structure. These may have a significant impact on a planet's environment as demonstrated by the appearance of ozone holes in the Earth's southern hemisphere. The depletion of ozone on the Earth occurs all over the planet, but, because polar vortices high in the atmosphere prevent air transport, ozone can deplete completely there and not be replenished from other parts of the planet. Although the Earth has vortices at both poles, chlorine clouds, which are essential to ozone destruction, only form at temperatures below about $-80\,°C$. Because these temperatures are rarely reached in the Arctic, the ozone holes have only formed in the Antarctic region. Polar vortices have also been observed on Venus [including a southern polar dipole (double eyed vortex) discovered by the recent Venus Express in 2006], and an enormous one on Saturn (Fig. 2.3).

In many cases, vortex filaments are not always visible because they are essentially "molecules" in a larger rotating flow. This is true for some approaches to atmospheric vortices where vortices are discrete elements of the barotropic or baroclinic flow. By taking the size of these filaments to zero and their numbers to infinity, we can describe very complex flows like super-rotation.

2.1.2 Aerodynamic Problems

In aerodynamic problems a great deal must be known both about the nature of the flow and the body under investigation and these explain the nature of, for example, wake vortices shed off the body's surface. These are often obtained by experiment. Issues such as the elastic nature of the body, its law of motion and deformation must be known; the conditions of flight, air currents, and turbulence present in the atmosphere a priori must also be taken into account. The medium is often considered to be unbounded, since this is a good approximation of flight far from the ground. Boundary conditions on the body must be specified. For a viscous medium, "no-slip" and "no-penetration" conditions are often used. No-slip means that the fluid molecules at the boundary (e.g., the wing or fuselage surface) are essentially stuck to the boundary with zero relative velocity with respect to it. In reality, this condition is not entirely true because molecules can often bounce along the surface, effectively "slipping" past the object, especially at high altitudes where pressure is low. The no-penetration boundary condition can be used instead which, as it sounds, means that the fluid cannot penetrate through the boundary. There also exists a class

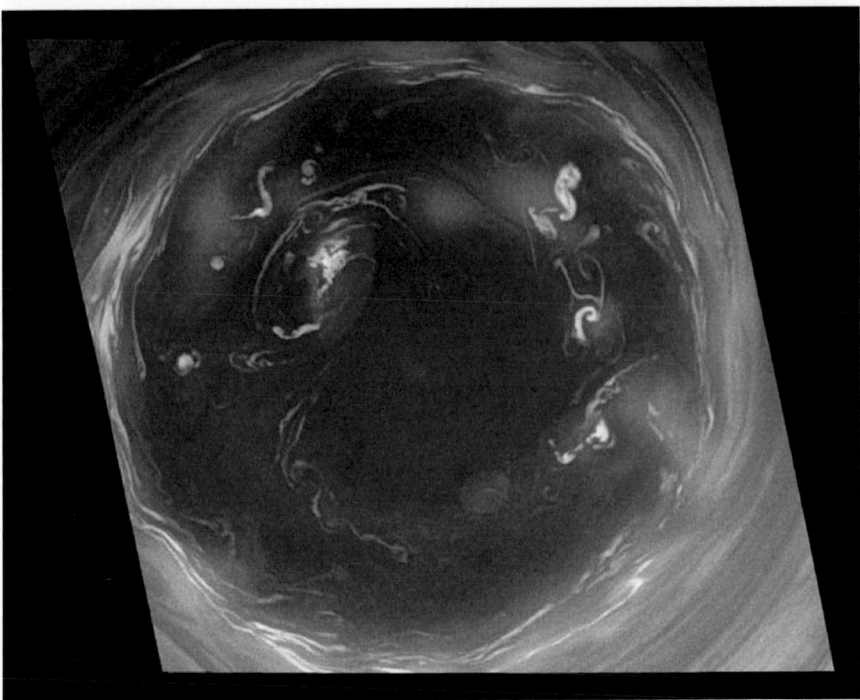

Fig. 2.3: This enormous polar vortex on Saturn was photographed by the Cassini probe. This vortex resembles a hurricane in some ways because of the clear air at the center, but it remains locked onto Saturn's pole rather than drifting around. Although it is not itself a filament, it can be broken up into a large number of infinitesimal vortex filaments. *Courtesy of NASA*

of interesting problems involving parachutes and paragliders with surfaces of thin fabrics (perhaps early aeroplanes also fit this category, certainly the Wright flyers, although powered, had surfaces of thin fabrics as did World War I fighters). These require that a law of cross-flow (since air can filtrate through the fabric) be specified by experimental determination of parameters.

Once all parameters and conditions are known, a vortex model must be applied. An instructive model is the method of discrete vortices where the body, such as an airfoil, is represented as a vortex surface of vortices moving through a medium and shedding vortices in its wake which then become "free" and move at the velocity of the medium. In this sense, both the body and its wake are represented as vortices. This is because flow past a body is not only translational and deformational but also rotational (where outside the region of the body flow is assumed to be irrotational). One can then represent the flow as a potential function, $\Phi(t,x,y,z)$, such that the velocity field is the gradient, $\mathbf{v} = \nabla \Phi$, and the potential obeys Laplace's equation, $\nabla^2 \Phi = 0$. In the case of an airfoil, for example, both the wing surface and the wake surface are replaced with a continuous vortex surface with the wing surface bound and the wake surface traveling with the fluid particles along stream lines [18]. This continuous vortex layer can then be used to calculate the velocity field based on the Biot–Savart law which gives the induced velocity for a vortex. These methods can also be applied to three-dimensional wings as well as blunt objects such as buildings or vehicles.

2.1.3 Bathtub Vortex

Perhaps the most common vortex encountered by almost anyone who uses a sink or bathtub is the aptly named *bathtub* vortex. The bathtub vortex has been the subject of some inquiry because it is commonly believed that, because of the Coriolis force, bathtubs in the northern hemisphere drain counterclockwise while those in the southern hemisphere drain clockwise. The issue turns out to be more complex.

A bathtub vortex forms as a fluid drains from a container through a hole. As the fluid depth lowers, rotational motion above the hole occurs, forming a vortex. In the absence of other factors, the Corolis force is the primary force for the rotational motion with a secondary flow in the boundary layer at the bottom surface of the bathtub or vessel. Indeed, because this secondary flow at the bottom overcomes the Coriolis force, bathtub vortices are observed to reverse direction at the end of the draining period. While some have contended that bathtubs or kitchen sinks drain in no particular direction in either hemisphere, they do so only because of mitigating factors. At the scale and speed of the water draining from the vessel, the Coriolis force is exceedingly weak, millions of times weaker than gravity. Under controlled experiment, in the Northern hemisphere, the vessel does drain counterclockwise, and, in the Southern hemisphere, clockwise [116], but, if the vessel is not perfectly symmetrical about the drain, asymmetries from the bottom and sidewalls of the vessel produce forces in excess of the Coriolis force, causing the vortex to rotate in another direction. This goes to show that boundary conditions are essential to most realistic studies of fluid motion including vortex models.

2.1.4 Deep Ocean Convection

The final hydrodynamical example of vortex filaments occurs in the deep ocean, far from land. Convection currents and eddies are a result of temperature gradients, and in the ocean these are responsible for a variety of behaviors from whirlpools to the Gulf Stream. In the deep ocean these can be quite visible because of differences in surface albedo (Fig. 2.4). Chapter 9 will explore deep ocean convection in a mean-field analysis of length scales in quasi-2D vortex ensembles.

Fig. 2.4: This eddy in Drake passage shows a strong surface albedo, making it visible in ordinary light. Again, this can be broken up into many infinitesimal filaments. *Courtesy of NASA*

2.2 Quantized Vortices

Quantized vortices occur in both type-II (high T_c) superconductors—where they are called London vortices—and in Bose–Einstein Condensates (BECs) and other superfluids. They are the subject of intense research, and, in the case of type-II superconductors, developing an understanding of these structures is of practical importance.

2.2.1 Bose–Einstein Condensates

In 1925 Einstein predicted, based an a paper sent to him by Satyendra Nath Bose, that the slowing of atoms in a cooling apparatus would condense into a singular quantum state now known as a Bose–Einstein condensate. Bose's original work was on the statistical mechanics of massless photons which Einstein then generalized to massive atoms.

Helium-4, with two protons, two electrons, and two neutrons, is one of the most studied bosons that forms a condensate, but, although liquid Helium appears to behave as a BEC, it is strongly interacting, hence it does not obey Bose–Einstein statistics precisely. Even so, in its superfluid phase it displays vortices. Moreover, these vortices are "quantized" in the sense that their circulations are integer multiples of a fixed constant. This is certainly not true of ordinary fluids which can have vortices of any circulation. Hence, BECs are part of a class of fluids known as "quantum" fluids because they display quantum behavior that cannot be explained with classical, Newtonian equations such as the Navier–Stokes. The liquid state of Helium-4 has two phases (1) a high temperature phase called Helium-I and (2) a low temperature phase called Helium-II. A phase transition, called the lambda-transition, separates the two. This transition, at $T_c = 2.172$ K, is the temperature at which Helium-4 begins to condense into a superfluid. Helium-I is a classical, Navier–Stokes fluid while Helium-II is a quantum fluid.

Helium-II contains two fluids, in fact, normal fluid and superfluid [16]. Each has a density, ρ_n and ρ_s, and velocity field, $\mathbf{v}_n$ and $\mathbf{v}_s$, for the normal and super fluids, respectively. The total density is $\rho = \rho_n + \rho_s$. The proportion of superfluid, ρ_s/ρ, approaches zero—no superfluid at all—as the temperature approaches a critical temperature T_c and unity as the temperature approaches $T = 0$ with about 90% superfluid at $T = 1.5$ K. Hence, the Navier–Stokes behavior is another complication in modeling Helium-II as a BEC because the normal fluid's behavior must be taken into account.

True Bose–Einstein condensation of dilute, weakly interacting atoms of Rubidium was achieved at the Joint Institute for Laboratory Astrophysics (JILA) and MIT in 1995 at a temperature about 170 nK. Through a combination of laser cooling and magnetic evaporation, these atoms were cooled sufficiently that they entered a condensed state such that they behaved as a single quantum atom. For this Eric Cornell, Carl Weiman, and Wolfgang Ketterle received the 2001 Nobel Prize in Physics. It represented the first time, since Einstein predicted it, that Bose–Einstein condensation could be observed, with a significant fraction of the atoms occupying the ground state. Since then research has expanded and BECs formed of several types of atoms.

The superfluid has some of the most unusual properties of any kind of matter. It has no entropy, is irrotational, has no viscosity, and infinite thermal conductivity (hence it cannot hold a temperature gradient, and any gradient is held by the normal fluid component). It can flow down a pipe with no resistance at all and escape through pores as small as 100 nm. In one experiment it is held in an open container within a larger, closed container. Anyone familiar with ordinary fluids will know that some fluids, like water, have a concave meniscus. The force that forms the

Helium-II "meniscus," however, remains greater than the force of gravity no matter how concave it is. Therefore, the Helium-II will creep up the sides of the container, forming a film called a Rollins film, and flowing into the larger container until all the fluid is at the same level. In another experiment, if a small tube is placed in a container of superfluid, the capillary effect is strong enough that a jet will emerge from the top in what is called the "fountain" effect.

Perhaps none of these effects, however, is as strange as the phenomenon that occurs when a cylinder of superfluid is rotated. Superfluid is irrotational, $\nabla \times \mathbf{v} = 0$, hence it cannot develop vortices as an ordinary fluid would. It must, however, represent angular momentum. Thus, when it is rotated beyond a critical angular velocity, $\Omega > \Omega_c$, it develops long defects a few angstroms thick, first one at slower velocities, then many at rapid rotation, that carry the angular momentum in a discrete form. These are quantized vortices. Given that the earliest interest in vortex filaments, by Onsager [117] and later Feynman [47], derived from their prediction, these vortices are the main reason why interest in vortex filaments remains strong, even after six decades of research. In the case of Navier–Stokes fluids, where vorticity is often continuous, vortex models are not as critical, but here the vortices are visible proof of the importance of discrete vortex filaments as fundamental features of nature, deriving from nature's most precise description—quantum mechanics.

2.2.2 Superconductors

Superconductors are to superfluids what electron plasmas are to ordinary fluids. Essentially, the vector potential $\mathbf{A}$ of a magnetic field such that $\mathbf{B} = \nabla \times \mathbf{A}$ is analogous to the velocity field, $\mathbf{v}$, such that $\omega = \nabla \times \mathbf{v}$. This analogy becomes more complete in magnetohydrodynamical systems.

Type-II superconductors or high T_c superconductors achieve superconductivity at higher temperatures than type-I by a mechanism that is not well understood. The highest T_c achieved for a compound that forms stoichiometrically (with definite formula) is 138 K and for a non-stoichiometric compound (with non-definite formula) 254 K or $-19\,^\circ$C, a mere cold day in temperate climates.

Superconducting materials all share particular properties which are a result of the thermodynamic phase. The first is exactly zero electrical resistance to low applied currents in the absence of a magnetic field. In an ordinary conductor electrons move through a heavy lattice. They scatter off the lattice, losing energy. Thus, the energy of the current is constantly dissipated. In a superconductor, however, electrons are bound together in pairs, called Cooper pairs, which, because they have integer spin, are bosons. Quantum mechanics creates an energy gap ΔE which is the minimum energy needed to excite the electron fluid. If the thermal energy of the lattice is less than the gap, no scattering occurs. Thus, the Cooper pair fluid is a charged superfluid or BEC. Indeed, a current in a superconducting ring can persist for years without significant dissipation. Thus, superconductors (and superfluids) are the closest phenomena in nature to perpetual motion machines.

2.2 Quantized Vortices

In an effect called the Meissner effect, magnetic fields below a critical strength only penetrate superconductors by about 100 nm before falling to zero. Fields above that strength, $\|\mathbf{H}\| > H_c$, destroy the superconductivity in type-I superconductors. Type-II superconductors have two critical field intensities. The first, H_{c1}, represents the intensity below which they react to the external field in the same way as Type-I superconductors. Above that critical value, however, there is a mixed state where the applied field penetrates the solid in the form of quantized flux lines. If the field is increased beyond a second critical value, H_{c2}, however, superconductivity is destroyed here as well. These quantized flux lines create some resistance, and, because they move within the solid, pinning them has become an important practical goal of superconductor research.

In 1957 Abrikosov [2] found an exact solution to the equations for type-II superconductors valid close to H_{c2}. His solution predicted that the order parameter forms a lattice of vortex lines. Abrikosov's result has been confirmed experimentally (Fig. 2.5), showing that the most common lattice is triangular (also shown in simulation, Fig. 1.1) although square lattices have now been found in some of the more exotic high-T_c superconductors. This prediction is one of the first predictions of emergent phenomena in a complex physical system, not just a macroscopic feature but a highly ordered one. This prediction and its resulting confirmation is so significant that Abrikosov shared the 2003 Nobel prize with Ginzburg. (The prolific Lev Landau was also a major contributor to this prediction but, besides having passed away 1968, had already won the prize in 1962 for his work on liquid helium.)

The Abrikosov lattice features prominently, not only in superconductors, but also in all areas of physics where parallel or nearly parallel vortex filaments interact. Hence, it is a fundamental feature of nature whenever this kind of vorticity appears. In superconductors, the Abrikosov lattice can be measured in a variety of ways, none simpler than the "decoration" experiment. In this experiment small paramagnetic particles are dusted onto the surface of the superconductor. Similar to spreading iron filings on a piece of paper above a magnet to see the force lines, the paramagnetic particles concentrate where the magnetic field is highest, over the flux lines, and the lattice becomes visible. For most ordinary high-T_c superconductors such as Pb and Nb, the lattice is regular and triangular (sometimes called hexagonal) [4].

The most important practical application for studying these vortices is vortex pinning. Because high-T_c superconductors (as opposed to type-I or low-T_c superconductors) are essential to many applications, e.g., MRI machines, which depend critically on their lack of resistance, it is desirable to pin the vortices so that they do not move around the superconductor. This prevents "flux-creep" which causes resistance and fixes the distance between the superconductor and the magnet. Flux-creep occurs when vortices move and dissipate energy. To keep them from moving, the vortices must be trapped in place by pinning. Pinning is created by structural defects in the superconductor and has a vast literature, but, ideally, the vortices should have a static Abrikosov configuration which requires that their interaction be strong enough to prevent fluctuations overtaking mutual repulsion. When fluctuations are too strong, the lattice is sometimes said to "melt" into a vortex liquid [120] which causes significant energy dissipation.

Fig. 2.5: An Abrikosov lattice of vortices in a type-II superconductor produced with magneto-optical imaging [60]. *source: Press material for the 2003 Nobel Prize in Physics* `http://nobelprize.org` *[Reprinted with permission]*

2.3 Plasmas

Several results in Chap. 7 relate to plasmas, specifically magnetohydrodynamics of electron plasmas.

In normal matter, the electromagnetic force creates structure in the form of atoms and molecules. This structure exists because the binding energy of the structures is greater than the ambient thermal energy. If the temperature exceeds the binding energy, these structures decompose. Atomic nuclei become free positively charged ions and electrons are no longer bound to them. This is called a plasma.

Plasmas are fairly common in the universe because so much of matter is bound up into stars, other hot matter, and their outflows. Thus plasma physics is one of the two major components of astrophysics (the other being gravitational physics). Plasmas are also common on Earth as lightning and, with modern technology, fluorescent lights and a variety of other common equipments such as arc welders. (Fire, once considered to be a separate state of matter by the ancients, is not hot enough to create a plasma. The glow is the result of electron excitation, but these electrons remain bound to their nuclei. Plasma televisions, on the other hand, work

2.3 Plasmas

by producing fluorescent light and, hence, actually do contain plasma.) Despite its general ubiquity, this fourth state of matter (or fifth if ultra-cold BECs are counted as the first as they should be) has only been known to science recently with the advent of the atomic theory of matter and the study of ionized gases in the late nineteenth and early twentieth centuries. Its name derives from the identical Greek word which means "jelly" and was first coined to describe blood plasma. Irving Langmuir, an American Nobel prize winning chemist, applied the term because electrons and ions being carried by a superheated fluid reminded him of the way blood plasma carries red and white blood cells. This interpretation is slightly confusing, though, since the electrons and ions *are* the plasma while blood plasma is certainly separate from the cells it carries. In any event, however, the name has stuck and become so common that the word "plasma" now evokes images of superheated matter and Solar explosions rather than the blood. Nevertheless, both are essential to human life [50].

2.3.1 Magnetohydrodynamics

A conducting fluid model may be applied to plasmas. This approach, pioneered by Hannes Alfvén in the 1940s and known as *magnetohydrodynamics* (MHD) combines Maxwell's equations with the Navier–Stokes to create a coupled model. Magnetohydrodynamical models have applications to Solar dynamics and the motion of the Earth's core and generation of the magnetic field. They have been applied numerically, for example, to predict the reversal of the Earth's magnetic poles (geomagnetic dipole reversal) assuming a dynamo model of the nickel-iron core (Fig. 2.6) [59]. This reversal occurs every few hundred thousand years and is now believed to be starting a reversal, the last having occurred 780,000 years ago. Because part of the core is a magnetized rotating fluid, vortex filaments are important features of the flow there in the same way that hydrodynamical currents can be described as vortex filaments in the ocean. Vortex filaments can also appear in the plasmasphere high above the Earth's surface.

In an MHD plasma, the plasma is bound to the magnetic flux lines, hence these flux lines, embedded in the plasma, determine the topology of the plasma. Vortices appear in the plasma as the charged particles orbit the flux lines. The velocity combined with the electric and magnetic fields satisfy Ohm's law,

$$\mathbf{E} + \mathbf{v} \times \mathbf{B} = 0, \tag{2.3}$$

which is known as the perfect conductivity equation or flux freezing equation. The latter comes from the observation that this equation implies a fixed topology, essentially, no flux lines can cross one another, and the flux lines move with the plasma, i.e., they are *frozen*. For example, it says that seemingly complex configurations of field lines can, over time, relax into very simple ones, while other equally complex ones will stay the same. This relates to knot theory, the original motivating

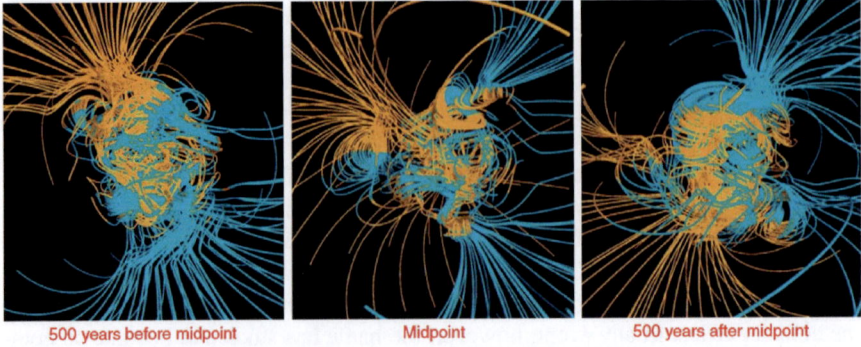

Fig. 2.6: This shows the magnetic flux lines emerging from the Earth's molten core progressing through a reversal. The dipole is in the process of reversing going from *left to right*. Vortex filaments are guided by the magnetic flux *lines* shown. Glatzmaier and Coe [58]. *Reprinted with permission from Elsevier*

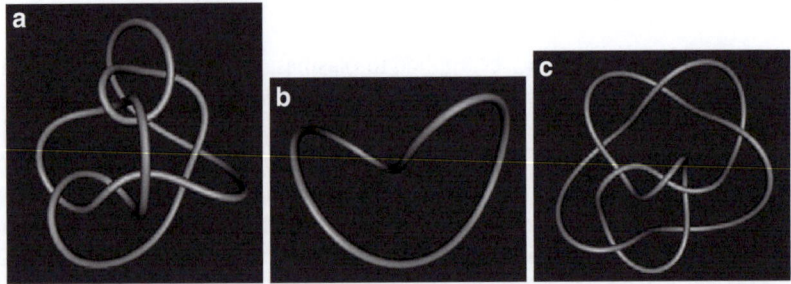

Fig. 2.7: (**a**, **b**) are equivalent under the frozen field equation, meaning that the first can be untangled to become the latter without any cutting or merging. Although it looks approximately as complex as **a**, the third (**c**) is an example of a "prime" knot and cannot be simplified further without cutting and/or merging. Vortex filaments typically behave like knots although cutting and reconnection are possible because of viscous effects. (Figures were generated with KnotPlot with **b** being achieved via a dynamical relaxation of **a**.) (**a**) A complex form of vortex equivalent to a ring, (**b**) the simplest form of the vortex ring or unknot, (**c**) the simplest form of a higher order vortex knot

factor in the study of vorticity (Fig. 2.7). The MHD frozen equation is a very strict requirement which is identical to the frozen field requirement in any inviscid fluid including BECs and Euler fluids and shows that viscosity is essential to merger and reconnection of vortices and flux lines.

2.3 Plasmas

2.3.2 Atmospheric Plasma

The plasmasphere is the inner part of the magnetosphere of the Earth, the region where the Earth's magnetic field dominates. It is just outside the upper ionosphere. When the Sun's rays strike the upper atmosphere, the ultraviolet rays excite the atoms to ionize. At the low densities of the upper atmosphere, plasma can form at very cold temperatures this way. Once the low mass electrons are freed from the heavier mass atomic nuclei, the electrons rise along the magnetic field lines leaving the positively charged ions below. This creates a powerful electric field as well. The densest area of plasma forms a torus around the Earth's middle (following the shape of the magnetosphere). While the plasmasphere is mostly uniform, irregularities in density can form plumes and prominences as well as vortices.

Lower down research involving plasmas in the ionosphere has grown out of a need to understand radio wave propagation. Here the physics is dominated by atmospheric and plasma physics rather than pure MHD. Indeed, ionospheric physics is particularly complex because it requires a *three* fluid model of neutral, ion, and electron fluids. These are acted on by the Earth's electric, magnetic, and gravitational fields. Alfvén vortices have been observed on Earth [42] as well as in the solar wind [5].

2.3.3 Solar Dynamics

Vortices have been observed in the solar atmosphere for decades with vortex direction in sun spots shown to relate to the solar rotation [129]. Quasi-2D and 3D MHD simulations have shown dense gas filaments form pinched bundles between magnetic loops, which produces a fine fiber structure perpendicular to the magnetic field direction [100]. Discrete statistical ensembles have also been applied to filamentary structures with applications to solar atmospherics [75]. (See also Fig. 2.8.) In recent years, thanks to the Solar Dynamics Observatory (SDO), Kelvin–Helmholtz instabilities (the "surfer's wave") have been observed in the sun [115]. These are another example of vortices in an MHD system.

2.3.4 Astrophysical Plasmas

Dust ion-acoustic shocks (both oscillatory and monotonic), Mach cones, dust voids, and vortices are all common nonlinear structures in dusty astrophysical plasmas. Dusty plasmas are essentially plasmas combined with charged interstellar dust grains [138]. Many astrophysical phenomena can now be simulated in the laboratory with lasers, generating small high energy plasmas, and dust ion-acoustic shocks have been observed in the lab [110]. Dust ion-acoustic (DIA) holes appear when warm ions follow a vortex ion distribution [137].

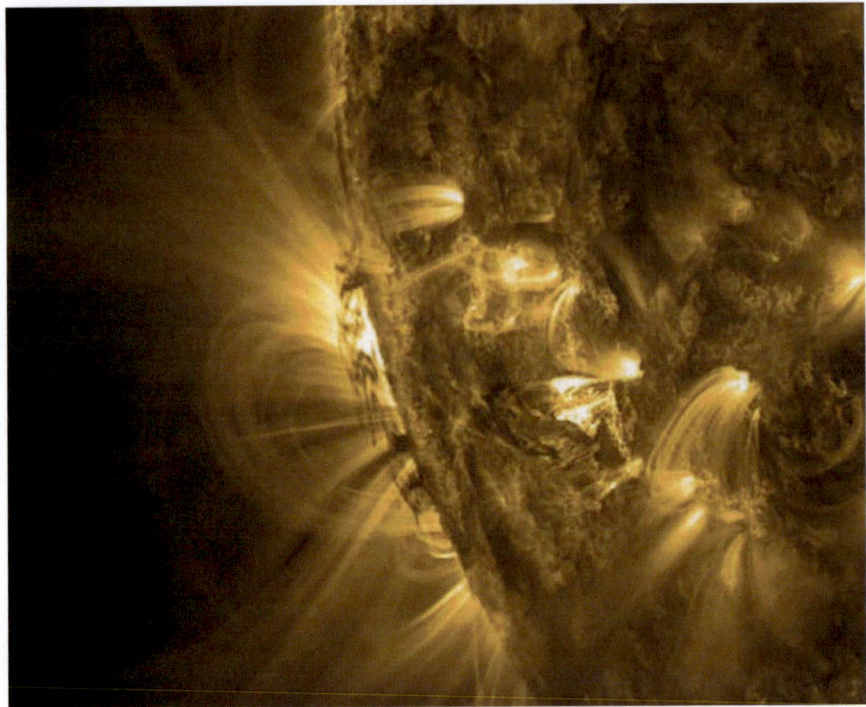

Fig. 2.8: Vortex filaments are common features of the solar atmosphere. Here we see several erupting out of the surface. These can be several times longer than the Earth. *Courtesy of NASA/SDO*

2.4 Concluding Remarks

In this chapter, we have seen how vortex filaments exist in many areas of physical reality. Indeed, anywhere where there is a fluidic-type structure whether it is water, air, liquid helium or dilute alkali atoms, Cooper pairs in superconductors, or ultra-hot plasmas, vortex filaments appear. In many cases, statistical mechanics only applies over very short time scales, and the behavior over longer time periods is highly dynamical. Nevertheless, statistical mechanics can give us insight into the emergent properties of these disparate phenomena that dynamical methods cannot.

Chapter 3
Statistical Mechanics

Statistical mechanics is a branch of physics that studies macroscopic systems from a microscopic point of view with the goal of understanding and predicting macroscopic phenomena from the properties of microscopic components [103]. Indeed, one of the underlying assumptions of physics is that all macroscopic phenomena derives from the action of microscopic components and that much of what is observed and measured in the natural world is the result of the statistical behavior of these components.

Statistical mechanics can be divided into two general areas of study: (1) systems in equilibrium and (2) systems not in equilibrium, and this book deals almost exclusively with the first one. The statistics of systems in equilibrium is often called statistical thermodynamics because it bridges the gap between molecular physics and classical thermodynamics. The motivation for statistical thermodynamics derives mainly from the weaknesses of thermodynamics which, while providing a great number of relations between different quantities such as heat capacity, pressure, energy, temperature, etc. and their derivatives, makes no physical interpretation for those quantities. In fact, thermodynamical relations would be valid even if atoms and molecules did not exist. Nor does it allow one to calculate physical properties separately; one can only relate them. To do these things, one needs statistical thermodynamics which requires the assumption of the existence of atoms and molecules.[1]

Statistical thermodynamics can be further broken into two parts (1) the study of systems where molecular interaction can be neglected or simplified and (2) the study of systems where the molecular interaction is of prime significance. The first kind is the simplest and was largely done up through the 1930s. Many of the great physicists of the twentieth century contributed to statistical mechanics including Einstein, Schrödinger [133], Pauli [119], and others. In the 1940s, 1950s, and 1960s, research moved on to the second kind of system, and a considerable number of

---

[1] Resistance to this assumption in the nineteenth and early twentieth centuries led to the severe criticism of the work of Ludwig Boltzmann which may have contributed to his committing suicide [91].

difficult problems remain, even for point particle systems, such as an exact solution to the one-component 2D Coulomb gas. The study of strong interactions in equilibrium of non-point particles, such as proteins, DNA, quantum particles over a stretch of time, and vortices (the main focus of this book) are particularly active areas of research with numerous practical applications to fields ranging from biomedicine to climatology.

3.1 History

Applying equilibrium statistical mechanics to large-scale quasi-2D flows (sometimes called 2.5D) remains a difficult subject to justify from first principles alone. To give a summary of the tradition of statistical mechanics we follow Bouvier in delineating this history into two phases, the first culminated in Gibbs' work of 1902 [53], after major works by Boltzmann and Maxwell, and the second started around 1920 when drastically reduced lattice spin models were proposed for the study of ferromagnetism. Gibbs' work emphasized the derivation of classical thermodynamics from the dynamics of atoms and molecules. He emphasized the invariance of phase-volume—Liouville theorem follows from the condition that the forces in the N-body problem depend only on positions of particles in the system and external bodies and possibly time—as a necessary condition for the ergodic hypothesis which is the cornerstone of his construction of the Gibbs' microcanonical ensemble. Sufficient condition for ergodicity is based on the extremely elusive metrical transitivity property. Ergodicity is mathematical tricky but intuitively it means that the averages of a single particle over time are the same as many particles over space. This is why statistical mechanics can apply to the state of a single particle over N units of time as well as N particles in space provided it is in equilibrium.

One of the key results of the first phase is the work of van der Waals, Maxwell and finally Ornstein which showed that combining a long-range attractive force with the short-range repulsion of a hard-core gas produced a non-monotonic (multivalued) pressure function of volume at fixed temperature, and justification of first order phase transitions. Everyday examples of first order transitions are the condensation of a gas into liquid and freezing of liquid into a solid. But even here, it was necessary to assume that the van der Waals' model satisfied the ergodic hypothesis because it remains to this day unprovable with no physically relevant exceptions. Many physicists have adopted the working principle that they should not have to prove the ergodic hypothesis in each new application of statistical mechanics. As a direct consequence of this we have on record more than a 100 years of highly successful theoretical work on the applications of equilibrium statistics to you-name-it fields from quantum to stellar systems. If physicists (and their mathematician-peers) had been held back by their inability to prove a theorem or ergodic hypothesis in a particular field, all these superbly successful theoretical work—with success measured, as any scientific model should be judged, by the qualitative and quantitative agreement between their predictions and experimental observations—would not have occurred.

Statistical mechanics entered its second phase with the proposal of the 1D Ising model by Lenz as a model for ferromagnetism. Two and three-dimensional versions of this model, together with numerous others such as the XY model, Heisenberg model, and Potts model, are characterized by extreme simplification of the interactions between spins at lattice sites. Such a paradigm shift—as indeed it turned out to be, for the Ising model was at the foundations of analytical and numerical treatments of phase transitions in lattice gas and lattice spin systems—can be safely termed a Minimalist Approach, because all the quantum and EM complexities in the interactions between electrons say have been swept away. Moreover, the lattice site values, often representing quantum spins, are here allowed to take very simple classical values, such as ± 1, or real values between $-m$ and m. This minimalist approach continues in many areas of statistical mechanics, including the study of vortex filaments where much of the internal structure and self-interference is swept away and tend to be justified by their testable predictions.

3.2 Ensembles

To study a system in statistical equilibrium, one must choose an ensemble. An ensemble defines how a large number of particles or objects interacts with its environment. While in ordinary mechanics, conserved quantities such as energy or momentum are normally discussed as if a system is always in isolation, in statistical mechanics we consider what happens when a system can exchange those quantities with its environment in a random but statistically known way. Which quantities are conserved and which are exchanged, determine the ensemble.

There are three main ensembles of statistical thermodynamics:

The first and most common ensemble is the canonical where an ensemble exchanges heat with its environment and temperature is fixed. This is typical of most laboratory setups. To keep volume, temperature, and the number of particles constant, the canonical ensemble is held within a rigid, conducting container where the container walls are an impassable barrier for the molecules within and without (Fig. 3.1). The molecules, however, can exchange heat through the container walls, and, thus, those within the container reach the same temperature as those without. This ensemble is the most frequently treated equilibrium ensemble and was the earliest to be studied.

The second ensemble is known as the "grand canonical ensemble." In this case, the formerly rigid container becomes porous and molecules are now allowed to enter and leave the ensemble (Fig. 3.2). Thus, in the grand canonical ensemble temperature is still fixed but the number of molecules fluctuates with the variations determined by a chemical potential. Many biological organisms fit into this ensemble because they have porous membranes that allow material to enter and leave. The chemical potential determines how much the number of particles fluctuates. A variation on the grand canonical is to allow the volume of the container

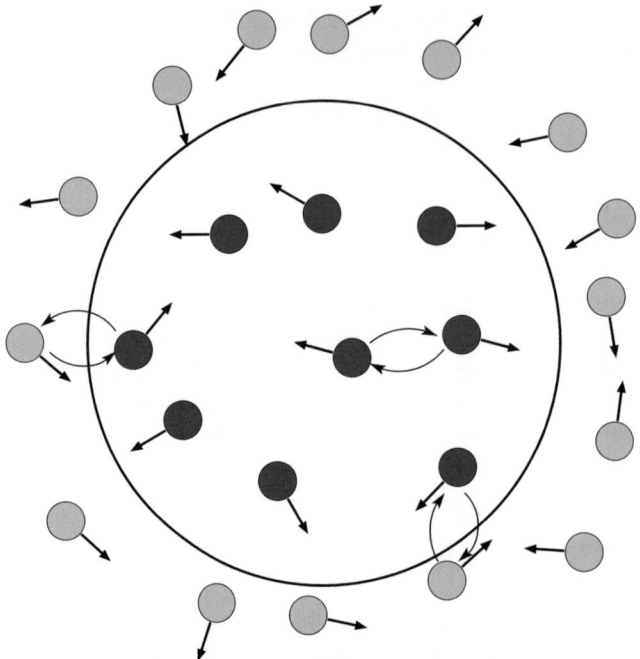

Fig. 3.1: The thermally conducting container allows energy to transfer to and from the bath, keeping temperature constant while energy fluctuates. Number of particles is still absolutely conserved

to fluctuate, by making it plastic, and keeping the number of particles fixed. In this case, the number of particles does not change but the volume each particle consumes does.

The third type of ensemble is the "microcanonical ensemble" where the energy is fixed by isolating the chamber from all outside influence. If the chamber from the canonical ensemble is completely non-conducting, it becomes a microcanonical ensemble (Fig. 3.3). The microcanonical ensemble is difficult to study in the laboratory where boundary conditions are almost always significant, but it applies to many natural systems that are effectively isolated such as globular clusters and systems of vortex filaments. Any number of other macroscopic properties besides energy can also be fixed in this ensemble.

The set of canonical ensembles, including the grand canonical, and the set of microcanonical ensembles constitute the two main approaches to modeling the statistics of physical systems. Surprisingly, they often generate identical predictions. When they do not, difficulties can occur that have only been discovered in the past few decades.

These difficulties are most often studied as non-equivalences between the canonical and microcanonical ensembles. When they are equivalent, as was assumed to

3.2 Ensembles

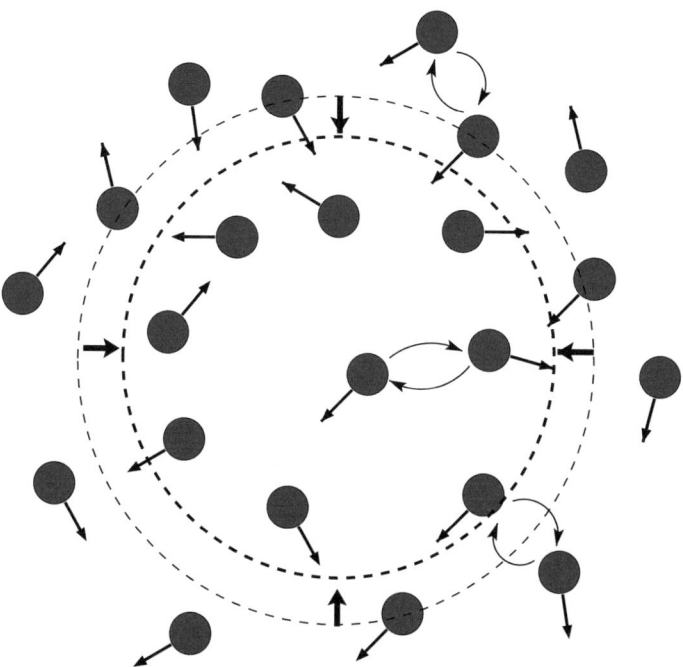

Fig. 3.2: The porous, plastic, thermally conducting container allows energy and molecules to transfer to and from the bath. To keep the number of particles constant, the volume of the "container" fluctuates

be always the case for a long time, the canonical ensembles offer a technically less difficult way to do calculations and make predictions in applications.

In contrast to the microcanonical approach [24, 32, 67, 68, 97, 106, 130] the extremal free energy in the canonical approach [81, 87, 89] is attained by balancing the internal energy U and the entropy S at any given heat bath temperature T (positive or negative in which case, we want the maximum free energy). Thus, at critical values T_c which generically have small absolute values, phase transitions occurs in which the extremal free energy is attained, not by the maximum entropy state, but rather by a lower entropy, extremal internal energy state.

When they are not equivalent, they describe different regimes of statistical physics. In particular, for vortex filaments and atmospheric systems, negative specific heats are sometimes predicted in the microcanonical ensembles but never in the canonical ensembles. Equivalence between the canonical and microcanonical ensembles is known to break down under certain conditions such as long-range Coulombic and logarithmic interactions [146], except in special cases with specific constraints and spectral degeneracy [67]. By a Fourier transform to the relevant orthonormal eigenfunctions or harmonics, made available on closed oriented manifolds by Hodge's theory, it can be shown that the logarithmic energy interactions of

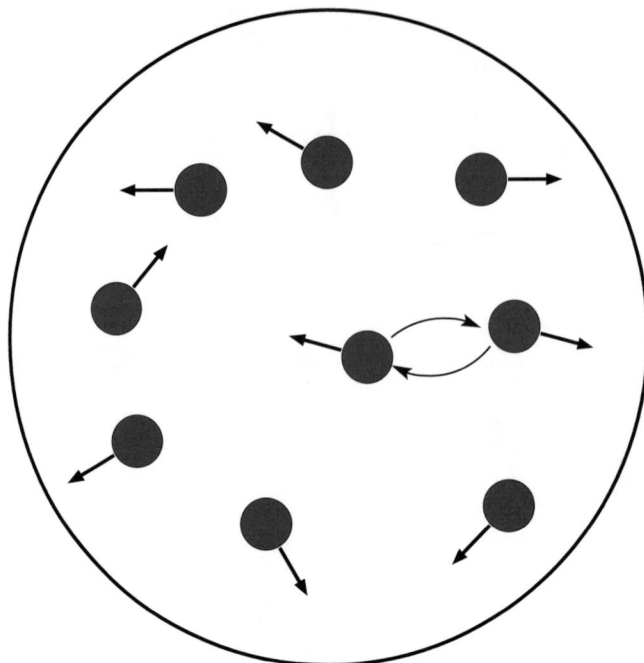

Fig. 3.3: Each water molecule has kinetic energy plus a van der Waals interaction with other molecules. The positions and velocities of all the molecules is a state of the system. No matter what the state, in this isolated system, energy and number of particles are absolutely conserved

the barotropic flow on the surface of a sphere, for instance, (which is a closed oriented manifold) can be expressed in diagonalized form with zero range [87]. Hence, the equivalence of the microcanonical and canonical ensembles in this special case [87].

Another application where the long-range logarithmic energy interactions of 2D fluid turbulence can be treated canonically, and where one can show equivalence between the microcanonical and canonical ensembles, is the inverse energy cascade to organized vortex crystals/lattice in a pure electron plasma trapped in a Penning-Malmberg trap. This application, sometimes referred to as a cold plasma, is very different from the situation for a hot plasma in a tokamak, where in cross-section, the plasma is supported on a free boundary domain, because the containing magnetic fields are designed to keep the hot plasma from touching the fixed toroidal walls of the device. In the hot plasma, Hodge's theory is naturally not available because the 2D flow domain in this case is essentially the whole unbounded plane. Thus, in the case of the hot plasmas, the microcanonical ensemble, being not equivalent to the canonical one, sometimes predicts negative specific heats, which have been considered as a possible source of tokamak plasma instabilities.

3.3 Assumptions

The purpose of equilibrium statistical mechanics is to calculate thermodynamic properties of "observables," i.e., expectations or averages of measurable quantities. It is a deep truth of physics in general that most of the things we measure are, in fact, averages of the random behavior of large numbers of microscopic entities. In classical statistical mechanics, the underlying assumption is that these macroscopic states are the result of myriad microscopic states that all generate the same average measurements. When we measure an observable such as pressure, we are not interested in the individual behavior of trillions of molecules striking our barometer, only the aggregate result. Statistical mechanics is a way of modeling all these trillions of individual behaviors at once in order to make predictions without knowing which microscopic states actually occur when we take our measurement. A microscopic state is also called a *microstate*. A macroscopic state is termed a *macrostate* and refers to thermodynamic quantities, e.g., a gas with a given pressure, temperature, and volume is in a macrostate and, no matter how much the microstates of individual particles change, the macrostate stays fixed as long as those parameters do not change. Often we consider macrostates to be collects or sets of microstates because many microstates can all give the same macrostate.

It is amazing, in a way, that statistical mechanics works as well as it does because so many of the mathematical assumptions are at odds with reality. First, statistical mechanics makes statistical predictions by averaging over all microstates even though only a small fraction actually occur in any given experiment. When we put a thermometer into a roast, after it has reached equilibrium, we are essentially "polling" the microstates of the molecules inside the roast by allowing them to transfer energy to and from the thermometer. We will never sample all possible states. Second, many statistical mechanical predictions are made in the limit of an infinite number of infinitesimal particles (necessary for calculating phase transitions such as melting), clearly at odds with a reality of finite sized, finite numbered atoms and molecules. Third, it assumes that systems are in equilibrium. This means that we assume that time has gone to infinity. Equilibrium is sometimes obvious. For example, a cup of hot tea left in a room for several hours will reach equilibrium with the temperature of the room. We know that as long as the temperature of the room stays constant (e.g., because of a thermostat) the tea will remain in that equilibrium state. At some point, however, the room will change temperature, either because the thermostat was changed or because somebody left a door or window open or perhaps the heater broke. Time is not actually infinite. Thus, when we define equilibrium statistical mechanics, we are assuming that the environment is constant and we implicitly assume a time frame for the experiment during which the environment is constant. It may seem strange to belabor that point when we all understand it intuitively, but, in less familiar cases, our intuition may fail us. For example, by the same logic, some system reach equilibrium in a few nanoseconds while the environment only stays constant for, say, a millisecond. In these cases, equilibrium statistics applies equally as well as in the case of a cup of tea in a room even though the supposedly infinite time we assumed is less than a millisecond. Hence, systems that, on their face, seem extremely dynamic may also be subject to equilibrium statistics.

What all this means is that, as with any model, equilibrium statistical mechanics has a range of validity, i.e., where (1) microstates that occur are a valid statistical sample (usually true if there are enough particles), (2) if the infinite particle limit is taken, the statistical behavior of the system does not depend on the number of particles (e.g., does not blow up), and (3) the time scale for reaching equilibrium is much less than the time scale at which the system breaks down. Assuming that all these hold, we can begin to calculate thermodynamic quantities that generate accurate predictions, and, in some cases, we can predict previously unknown microscopic processes underlying macroscopic phenomena. For instance, a more complete study of the dynamics of planetary atmospheres than that provided by equilibrium statistical mechanics will depend on new advances being made in nonequilibrium statistical physics [25, 84, 98, 99].

3.4 Methods

There are two well-known approaches to deriving thermodynamic quantities in statistical mechanics. The first is the calculus of variations, involving functional derivatives to find the most-probable states, and the second is via iterated or functional integration over states. We will apply the calculus of variations to a vortex problem in Chap. 7 and the functional integration approach in Chap. 9. They are often equivalent but, depending on the result desired, one approach may be quicker and more elegant than the other. Both approaches center on a description of the given system in terms of likelihoods (or probabilities) of particle states. A density function, which we refer to as $f(\chi)$, describes the density of particles at any point in the phase-space $\chi \in D$ and particles are assumed to be self-contained with no possibility of self-interference. For particles in three space, for example, $\chi \in \mathbb{R}^6$ and may be split into a position vector $\mathbf{r} = (x,y,z)$ and velocity vector $\mathbf{c} = (u,v,w)$, with $f(\mathbf{r},\mathbf{c})$ describing the likelihood of a particle having the given position and velocity. Often, we also refer to the positional density function (because velocity should be integrated out at the end of most calculations), $\rho(\mathbf{r}) = \int d\mathbf{c} f(\mathbf{r},\mathbf{c})$. We will first describe the approach via calculus of variations.

3.4.1 The Variational Approach

The variational approach can be used to derive one of the most basic formulas in atmospheric science: Given uniform temperature, the pressure of a classical gas in a uniform gravitational field with acceleration g decreases with height h by the barometric formula:

3.4 Methods

$$P(h) = P(0)\exp{-mgh/k_B T},$$

given that $P(h) = \rho(h)k_B T/m$ and $\rho(h)$ is density. In this section, we are going to tackle a slightly different but similar problem.

Suppose we are given a system with energy functional, $E[f]$, and asked to derive the most likely equilibrium density function f_{max}, such that the energy, E, and particle number, $N[f] = \int d\chi\, f(\chi)$, are fixed. (Additional conserved quantities such as pressure, volume, and angular momentum may also be added, but, we will come to these in more specific examples later.) According to equilibrium thermodynamics, a system in isolation reaches its equilibrium when its entropy is maximized. Therefore, we would like to find the point, f_{max}, at which the entropy achieves its maximum.

Gibbs gives the entropy as,

$$S[f] = -k_B \int_D d\chi\, f(\chi)\log f(\chi), \tag{3.1}$$

where k_B is Boltzmann's constant. From this point forward, we let $k_B = 1$ and drop it from all further derivations. The meaning of the entropy has a variety of interpretations, but the main one in statistical thermodynamics is that it is the log of the number of microstates in a given macrostate. Thus, it is a measure of uncertainty. The more microstates that are consistent with our macroscopic measurements, the greater our uncertainty about the microstate of the system. A major difference between the variational approach and the integral approach to follow is that the variational approach explicitly maximizes the entropy, while the integral approach does so only implicitly.

A necessary (but not sufficient) condition is that the variation of S is zero, $\delta S = 0$, at f_{max}, subject to the constraints that energy and particle number are conserved:

$$\delta S = 0 \tag{3.2}$$
$$E[f] = E_0 \tag{3.3}$$
$$N[f] = N_0. \tag{3.4}$$

Once this is solved, a physical argument usually determines whether f_{max} is at a global maximum of S (as opposed to a local maximum, local minimum, or saddle point). We solve the constrained maximization problem by the method of Lagrange multipliers. Each conserved quantity has its own Lagrange multiplier. The density, f_{max}, is the solution to the unconstrained equation,

$$\delta S - \beta\,\delta E - \nu\,\delta N = 0, \tag{3.5}$$

where particular choices of β and ν fix the energy and particle number at the required values. Solving for f_{max} is easier said than done, however. For any but the most trivial systems (with some notable exceptions covered in a later chapter), one must rely on approximation methods such as perturbations for weakly interacting particles, numerical methods, or mean-field approaches, which, nevertheless, may provide excellent predictions.

The variational procedure is similar for a fixed temperature system (i.e., the "canonical" approach as opposed to the fixed energy, "microcanonical" approach just described). In this case, we still must maximize the entropy but, because the energy is not fixed, we minimize the energy also. Therefore, for our fixed volume system, we are obliged to find the minimum of the, e.g., Helmholtz free energy, $F = E - TS$, (where the temperature, T, is merely another Lagrange multiplier), with the variation,

$$\delta E - T \delta S + \nu \delta N = 0, \qquad (3.6)$$

where T is given. Something odd about the variational approach though is that we could have as easily minimized the non-dimensional free energy, $\beta F = \beta E - S$, where $\beta = 1/T$, in which case the variational equations are identical. This means that the canonical and microcanonical ensembles are identical! The only difference is that, for fixed energy, the inverse temperature, β, may be positive or negative. In general, we require β to be positive in the fixed temperature case. (Negative absolute temperature heat bath systems have been studied in nuclear spin systems and, to be realistic, require an upper bound on energy [125].) Also, note that if pressure, p, rather than volume, V, were conserved here, we would conserve the enthalpy, $H = E + pV$, rather than energy alone or, alternatively, minimize the Gibbs free energy $G = E + pV - TS$.

Even if temperature is positive, that the fixed temperature and fixed energy cases are identical does not imply, as was once believed [133], that the "true" fixed energy and fixed temperature systems are always equivalent. Indeed, this is one of those cases, mentioned above, where our assumptions about the validity of our statistical mechanical model can be violated. Statistical mechanics describes the behavior of large numbers of particles but always a finite number. The variational approach, as given here, assumes a single, smooth density function, f, can represent all these particles as if they were infinitesimal. This function, however, only emerges from taking the limit as the number of particles approaches infinity (and scaling down their individual properties if necessary to maintain conservation of mass, charge, momentum, etc.). In some cases, however, this limit (called the thermodynamic limit) does not exist in the fixed temperature regime because of uncontrollable energy fluctuations (sometimes referred to as ultraviolet catastrophe, a reference to Planck's hot-body radiation problem which led to the invention of quantum mechanics). Therefore, while the above variational formula (3.6) applies to all fixed energy systems, it only applies to fixed temperature systems where the thermodynamic limit exists. When it does not exist the continuum variational approach only describes the microcanonical ensemble. To model the canonical ensemble, one must either (1) replace the variational formula with a large set of equations, one for each particle, (2) take a "non-extensive" limit in which coupling constants are scaled to force the limit exist (this is not the same as renormalization where coupling constants have to be scaled in both ensembles), or (3) take the integral approach.

3.4 Methods

We will now present a small example that will both illustrate the variational approach and give a first glimpse at vortex filaments. In Chap. 7, we will go through a derivation of some of our results on negative specific heat using the variational approach. Let a non-conducting, perfectly rigid drum, far from any outside influences, have radius R and height L and be filled with an ideal gas, and let the drum and gas be rotating at a fixed speed. The energy is fixed at E_0, the particle mass at mN_0, and the angular momentum, pointing in the positive z direction along the drum's central axis, is $I_0\hat{\mathbf{z}}$. The system's conserved quantities are given by the functionals,

$$E = \int d^6\chi \, f(\chi) \frac{1}{2} m|\mathbf{c}|^2$$

$$I = \int d^6\chi \, f(\chi) m(xv - uy)$$

$$N = \int d^6\chi \, f(\chi),$$

and the most likely density, $f_{\max}$, is the solution to the equation,

$$\delta S - \beta \delta E - \mu' \delta I - \nu \delta N = 0, \tag{3.7}$$

where S, given by (3.1), achieves its global maximum. Note that we threw in angular momentum as another conserved quantity. You can add as many conserved quantities as needed with a Lagrange multiplier for each. Applying the variations and (3.7) becomes

$$\int d^6\chi \, \delta f \left[\log f + 1 + \beta \frac{1}{2} m|\mathbf{c}|^2 + \mu' m(xv - uy) + \nu \right] = 0, \tag{3.8}$$

and, since δf is arbitrary, the integrand must also be zero,

$$\log f_{\max} + 1 + \beta \frac{1}{2} m|\mathbf{c}|^2 + \mu' m(xv - uy) + \nu = 0, \tag{3.9}$$

or, solving for the density, we have the Boltzmann distribution,

$$f_{\max}(\chi) = \exp\left(-\beta \frac{1}{2} m|\mathbf{c}|^2 - m\mu'(xv - uy) - \nu - 1\right), \tag{3.10}$$

as the most likely density. Integrating over the velocities (assuming positive temperature), the positional density is

$$\rho(\mathbf{r}) = A \exp\left[\frac{\mu'^2 m(x^2 + y^2)}{2\beta}\right], \tag{3.11}$$

where

$$A = \left(\frac{2\pi}{\beta m}\right)^{3/2} \exp[-(\nu+1)]. \tag{3.12}$$

Replacing $\exp[-(\nu+1)]$ with the appropriate normalization so that we have a total density of N_0 and converting to polar coordinates, the positional density becomes

$$\rho(r) = B \exp\left[\frac{m\mu'^2 r^2}{2\beta}\right], \tag{3.13}$$

where

$$B = \frac{Nm\mu'^2}{2\pi\beta L(s-1)} \tag{3.14}$$

and $s = \exp[m\mu'^2 R^2/(2\beta)]$. As expected, the drum acts like a centrifuge. Density increases exponentially outward, perpendicular to the axis of rotation.

Finally, we need to check that the Lagrange multiplier, β, is inversely proportional to the average kinetic energy of a single particle. The equipartition theorem states that

$$\frac{\langle E \rangle}{N_0} = \frac{\int d^3\mathbf{c}\, f(\chi)\frac{1}{2}m|\mathbf{c}|^2}{\int d^3\mathbf{c}\, f(\chi)}, \tag{3.15}$$

ought to give $\frac{3}{2\beta}$ for an ideal gas, but, in this case, we have chosen our multipliers inappropriately, and we find that the calculation is incorrect. Because velocity appears linearly in the expression for angular momentum, the multiplier for that conserved quantity should be $\sqrt{\beta}\mu$ instead of μ'. Once this substitution is made, (3.15) evaluates to $\frac{3}{2\beta}$, and we may pronounce β to be the true inverse temperature. The additional advantage is that we remove β entirely from the positional density and can easily give the formula for energy, $E = \frac{3N_0}{2\beta}$. The angular momentum, by a similar method, is

$$I = \frac{N_0}{\sqrt{\beta}}\left[\frac{R^2\mu s}{1-s} + 2/\mu\right]. \tag{3.16}$$

Thus, we have gone through a toy model to illustrate the method of variations but also to show some of the pitfalls. In interpreting Lagrange multipliers, the energy multiplier cannot blithely be regarded as inverse temperature without proving the equipartition theorem.

A couple points to make about this experiment (which can be done in a lab, although friction with the container restricts any results to short time scales between the formation of the equilibrium state and its dissipation). First, integrating angular momentum, $\mathbf{L} = m\mathbf{r} \times \mathbf{c}$, over velocity in the example above gives a standard statistical angular momentum expression for rotating fluids centered at the origin: $m(x^2+y^2)$. This expression will represent angular momentum in all future calculations.

The second, more profound point is that the low density at the center of the drum is none other than a vortex filament, albeit not one that follows the Euler or Navier–Stokes equations. This ideal filament displays one interesting quality of traditional vortex filaments though. Angular momentum has the opposite effect on it as on ordinary matter. Because filaments are "holes" in fluids, they gravitate toward the center under rotation with a force proportional to their distance from the origin almost like negative mass. Of course, anyone can see this by stirring a cup of tea. This effect is particularly important in deep ocean and atmospheric convection.

3.4.2 The Integral Approach

The integral approach to equilibrium statistical mechanics eliminates the necessity of using Gibbs' definition of entropy which is not always easy or appropriate. It is the method of choice when considering systems with a finite number of degrees of freedom and, as mentioned above, the two approaches can give different answers if the thermodynamic limit does not exist because of ultraviolet catastrophe. The integral approach has roots in probability theory of distributions, but statistical mechanics has developed its own nomenclature that is more appropriate when considering particle configurations in equilibrium.

In probability theory, we have an event space, $\mathscr{E}$, and an element of that space is an event, $E \in \mathscr{E}$. The word "event" implies that we are measuring the probability of something happening, which is common in forecasting. In statistical mechanics, however, our events are configurations of particles; hence, the event space is called the *configuration* space (sometimes also called the *phase-space* or *state space*), $\mathscr{A}$, and an element in that space as a configuration, $A \in \mathscr{A}$. The configuration space also has a size or cardinality that is given by a function, $Z(\beta)$ or $\Omega(E)$, for the canonical/grand canonical and microcanonical ensembles, respectively. The function is usually called the *partition function* but sometimes the *statistical weight* or *normalization*. The configuration space together with a likelihood function, $\lambda : \mathscr{A} \to \mathbb{R}^+$, on that space is called a *statistical ensemble*.

3.4.2.1 Canonical Ensemble

As explained above, the most common statistical ensemble is the *canonical* or fixed temperature ensemble. For a system with N distinguishable particles each with position $\mathbf{r}_i$ and velocity $\mathbf{c}_i$, potential energy $\phi[\mathbf{r}_1, \ldots, \mathbf{r}_N]$ (V is commonly used for potential energy, but we reserve this symbol for volume), kinetic energy $K[\mathbf{c}_1, \ldots, \mathbf{c}_N]$, and temperature β, the probability density of a microstate is given by

$$p = \frac{\lambda}{Z}, \qquad (3.17)$$

where

$$\lambda = \exp[-\beta(K+\phi)], \qquad (3.18)$$

and

$$Z = \int d\mathbf{c}_1 \ldots d\mathbf{c}_N d\mathbf{r}_1 \ldots d\mathbf{r}_N \, \lambda, \qquad (3.19)$$

is the partition function. The ideal gas law, $PV = nRT$, may be derived from the case where $\phi = 0$ and $K = \frac{1}{2}mc_i^2$, and, in the case of non-zero potentials such as van der Waals, nonlinear corrections to the ideal gas law may also be derived. Since this is not a book about general statistical mechanics, we will not derive these explicitly but move onto the next ensemble, the *microcanonical* or fixed energy ensemble.

3.4.2.2 Microcanonical Ensemble

Until Thirring's 1970 paper [142], the canonical ensemble and its extension, the grand canonical ensemble which allows the number of particles to change as well, were considered to represent all statistical mechanics, hence the term canonical. A conflict, however, arose when some systems, in particular gravitational systems such as globular clusters, were found to have negative specific heat [95], which was thought to be impossible.

Long before this, Erwin Schrödinger had "proved" that all statistical systems have positive specific heat using the canonical ensemble [134]. This proof is straightforward: the specific heat of a statistical ensemble is the rate of change of temperature with respect to mean energy assuming either constant volume,

$$c_v = \left.\frac{\partial \langle E \rangle}{\partial T}\right|_V, \qquad (3.20)$$

or constant pressure,

$$c_p = \left.\frac{\partial \langle E \rangle}{\partial T}\right|_P. \qquad (3.21)$$

These are usually with respect to some mass of the ensemble of course such as 1 gram. In the canonical ensemble, $\beta = 1/k_B T > 0$ where k_B is Boltzmann's constant. In this book, we will always assume units where $k_B = 1$. Therefore, with respect to β the specific heat is

$$c_v = -\beta^2 \left.\frac{\partial \langle E \rangle}{\partial \beta}\right|_V, \qquad (3.22)$$

and likewise for c_p. The mean energy of the ensemble is the energy of the most-probable macrostate and has the formula,

$$\langle E \rangle = \frac{1}{Z} \int d\mathbf{c}_1 \ldots d\mathbf{c}_N d\mathbf{r}_1 \ldots d\mathbf{r}_N \, E \exp[-\beta E], \qquad (3.23)$$

where $E = K + \phi$. If we assume that Z is finite (not always true), then

$$-\langle E \rangle = \frac{\partial \log Z}{\partial \beta}. \qquad (3.24)$$

Since Z has a positive integrand it must also be positive, so the logarithm does not pose a problem and it defines the entropy of the canonical ensemble,

3.4 Methods

$$S = \log Z. \tag{3.25}$$

The specific heat is

$$c_x = \beta^2 \left.\frac{\partial^2 \log Z}{\partial \beta^2}\right|_X, \tag{3.26}$$

where $x = v, p$ and $X = V, P$. This is the second moment of the probability distribution,

$$c_x = \left(\beta^2 \frac{1}{Z} \int d\mathbf{c}_1 \ldots d\mathbf{c}_N d\mathbf{r}_1 \ldots d\mathbf{r}_N E^2 \exp[-\beta E]\right)_X, \tag{3.27}$$

which is positive even if the energy takes on negative values. Thus, the specific heat of the fixed temperature ensemble is always positive.

This seemed to confirm that negative specific heat is impossible because the variational approach indicates that the fixed energy and fixed temperature ensembles are identical. The variational approach requires a continuous density function and in that case a single Lagrange multiplier, β, corresponds to a single fixed energy value, E. Therefore, in the variational approach, fixed temperature means fixed energy as well. As mentioned, however, the "proof" that specific heat is always positive relies on the partition function being well behaved as the number of particles goes to infinity, $N \to \infty$, which is necessary to arrive at the variational solution. The partition function, however, is not well behaved when the pair body potential has an infinite range, e.g., $\phi(r_1, r_2; q_1, q_2) = -q_1 q_2 \log(r_1 - r_2)$ for a one-dimensional Coulomb gas. The potential scales with the square of the number of particles because every particle interacts with every other particle no matter how far apart they are. On the other hand, the kinetic energy, e.g., $K(c_1; m_1) = \frac{1}{2} m_1 c_1^2$, is a self-energy and scales linearly with the number of particles. Therefore, the partition function *per particle*, Z/N, blows up because the contribution of the potential energy does not scale with N but N^2 and, if we instead consider the partition function per particle pair, Z/N^2, we lose the kinetic energy to zero. Neither approach leads to the variational solution because the ensemble is wrong. As we will show later in the book, there are ways around this blow up (requiring a special scaling of the energy), but they are not applicable to every system state and certainly not those with negative specific heat.

Unlike neutral molecules and atoms whose interactions have a definite distance cut-off, because stars, charged particles, and vortex filaments interact to infinite distance, they may have negative specific heat and even negative temperatures, and, if one wants to consider them in a statistical ensemble, one must use the microcanonical likelihood:

$$\lambda = \delta(E_0 - E), \tag{3.28}$$

where δ is the Dirac-delta function, and, to distinguish it from the canonical partition function, we give the microcanonial partition function the symbol Ω.

The microcanonical likelihood has no mention of temperature in the ensemble itself, and, in fact, a microcanonical ensemble may have any temperature for a given energy. It does, however, have an expected temperature which can be obtained from the entropy,

$$\frac{1}{\langle T \rangle} = \frac{\partial S}{\partial E_0}, \tag{3.29}$$

where $S = \log \Omega$, which is the same as the formula for temperature in the canonical ensemble, but, in this case, the expected energy is the fixed energy value, $\langle E \rangle = E_0$. The microcanonical ensemble is less common in statistical mechanics because it is harder to use, and, in most cases, it is equivalent to the canonical. This equivalence comes from the Fourier transform of the Dirac-delta function,

$$\delta(E_0 - E) = \int_{-\infty}^{\infty} \frac{d\eta}{2\pi} e^{i\eta(E_0 - E)}, \tag{3.30}$$

which under the change of variables $\beta = i\eta - \beta_0$ becomes

$$\delta(E_0 - E) = \int_{\beta_0 - i\infty}^{\beta_0 + i\infty} \frac{d\beta}{2\pi i} e^{\beta(E_0 - E)}, \tag{3.31}$$

where $\beta_0 = 1/\langle T \rangle$. Assuming that E increases linearly with the number of particles, N, steepest descent implies that for large N,

$$\int_{\beta_0 - i\infty}^{\beta_0 + i\infty} \frac{d\beta}{2\pi i} e^{\beta(E_0 - E)} \approx C e^{-\beta_0 E}, \tag{3.32}$$

where $C = \exp[\beta_0 E_0]$ is a constant. In the thermodynamic limit, $N \to \infty$, with appropriate normalization to keep the result from blowing up, this approximation becomes equivalence. Again, the existence of this limit depends on E increasing linearly with N.

3.5 Statistical and Fluid Mechanics

The field of fluid mechanics is traditionally concerned with deterministic equations that model the flow of fluids given boundary conditions. The properties of fluids, however, may only be describable in a statistical sense, while exact configuration, composition, momentum, and energy are only known probabilistically. The philosophy behind applying statistical mechanics to fluids is that it can capture the broad features of a system without needing to understand the details. Hence, we can make analogies between one system and another that are superficially very different, e.g., between phase transitions in Ising models and those in vortex fluids. In essence, statistical mechanics has always been about gaining a broad understanding of the connection between the emergent phenomena we observe and the microscopic interactions of myriad particles and, critically, *classifying* the phenomena irrespective of the underlying system. In any field where emergent phenomena is important, including fluid mechanics, statistical mechanics has an important role to play. In the real world there are no truly equilibrium statistical phenomena. Nonetheless, for many situations where the relaxation occurs at a faster time

3.5 Statistical and Fluid Mechanics

scale than the dissipation, equilibrium statistical mechanics have provided accurate predictions of forced-damped systems out of thermodynamic equilibrium [95, 97]. Within the equilibrium statistical mechanics framework, following the classical works of Onsager [117], T. D. Lee, Kraichnan [81], and Leith [83], several theories have been proposed to address quasi-2D turbulent relaxation phenomena in macroscopic fluids in a wide variety of geometries. The earliest successful class of theories after Onsager, Lee and Kraichnan are based on the point vortex model [93]. The successful Miller–Robert–Sommeria theory [106, 130] and the Majda–Turkington theory [146] which is based additionally on prior statistics of the forcing or small scales [97] are examples of the microcanonical approach. They are based on maximal entropy rearrangements of non-overlapping vortex parcels where some or all moments of vorticity are conserved. In order to derive useful relations between the macroscopic quantities, the (microcanonical) statistical ensembles in these theories are often approached from the mean-field approximation. Large Deviation Principles have been used to prove that the mean-field PDE are exact [130, 146]. Subtly different predictions for the most-probable states were found in certain 2D turbulent flows [97, 130], especially in the case of 2D turbulent flows that are forced at small scales, partly because the different approaches assume distinct a priori statistics and also conserve different number of vorticity moments. A recent paper showed that the inclusion of the quartic vorticity moment into the equilibrium statistical mechanics formulation has subtle consequences at the level of the microstate [98].

Chapter 4
Parallel Filaments

The previous three chapters have provided the background and motivation for studying vortex statistics. Now we begin to look at ensembles, starting with the simplest, parallel vortex filaments. In the literature, one will see the words "vortex line" used as often as "vortex filament." Technically, these are different entities. In his seminal treatise on hydrodynamics, Horace Lamb defines the properties of vortices drawing on the pioneering work of the nineteenth century scientists Stokes, Helmholtz, Kelvin, and others [82]. A vortex line is described as the axis of rotation of a fluid. In particular, a vortex line is the axis of rotation of a rotating tube of fluid or other substance. Lamb distinguishes a vortex line from a filament saying that the former is the axis while the latter is the rotating fluid contained about the axis, i.e., the vortex itself. In this book, however, because we do not go into the behavior of the vortex core but consider a vortex to be a one-dimensional, string-like molecule, the two terms are used interchangeably.

Parallel filaments are the simplest model of filament vorticity ensembles because they can be represented as two-dimensional points. These simple ensembles have captivated the interest of researchers for decades because they display an unusual empirical phenomenon: negative temperature states. Negative temperature states are believed to be important in the formation of large regions of vorticity such as hurricanes where like-signed vortices (those rotating in the same direction), which normally repel, combine to form larger and larger vortices. The inversion of repulsion is an effect of negative temperature states on logarithmically interacting bodies because $\beta \log r_{jk} = -\beta \log 1/r_{jk}$ where r_{jk} is the distance between vortex j and vortex k. Hence, when the inverse temperature Lagrange multiplier $\beta = 1/T$ couples to the energy (see Eq. 1.3) the negative sign inverts the distance in the log. Parallel filaments are also important in the study of neutral plasmas, where filaments are artificially aligned via a magnetic field and which also show negative temperature states. (This connects to what we said in the conclusion of the previous chapter about disparate systems sharing common emergent phenomena.)

© Springer Science+Business Media New York 2014
T.D. Andersen, C.C. Lim, *Introduction to Vortex Filaments in Equilibrium*,
Springer Monographs in Mathematics, DOI 10.1007/978-1-4939-1938-3_4

4.1 The Point Vortex Gas

Lars Onsager initiated the study of statistical ensembles of parallel vortex filaments by inventing the 2D point vortex or "Onsager" gas [113, 117]. (See Fig. 4.1 for a picture of a vortex gas state.) Onsager took this statistical approach to understand an aspect of turbulence, specifically how large-scale meteorology structures form from clusters of smaller vortices or eddies even though basic hydrodynamics states that like-signed vortices repel one another like electric charges. Traditional hydrodynamics failed to provide an answer. What could possibly prompt such a reversal of what appears to be cardinal physical law? After some clever mathematical and physical arguments, Onsager's explanation was that ensembles of like-signed vortices have negative temperature states, which force energies to higher and higher levels. Negative temperature states turn out to be important in explaining the forma-

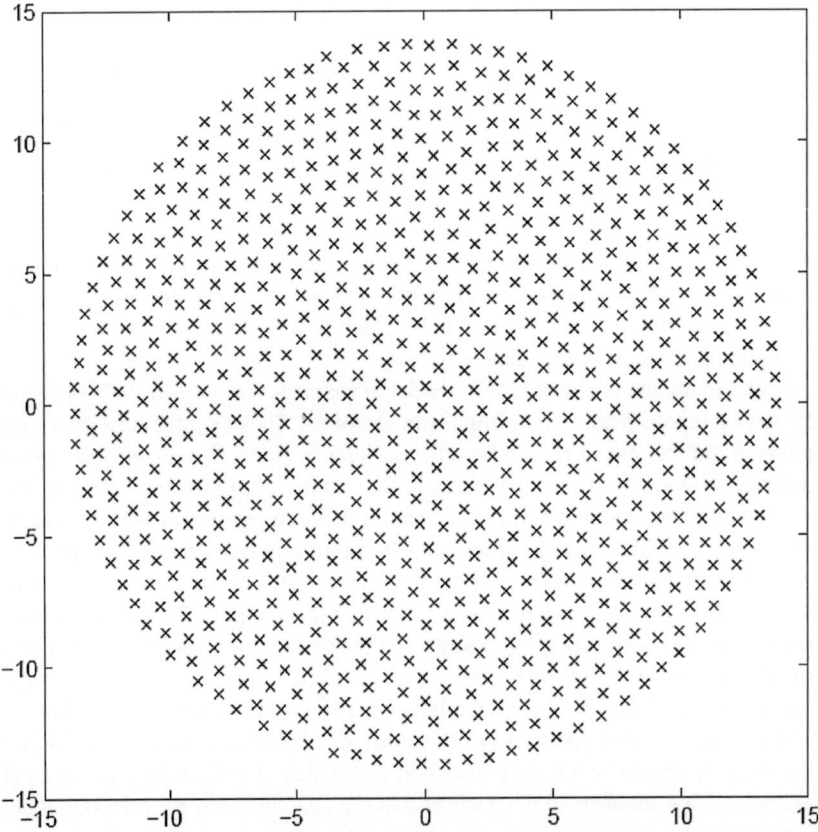

Fig. 4.1: The point vortex gas is a gas of interacting filaments. Here the points are represented as "x"s to make them easier to see but they are infinitesimally thin points of vorticity in the 2D plane [88]

4.1 The Point Vortex Gas

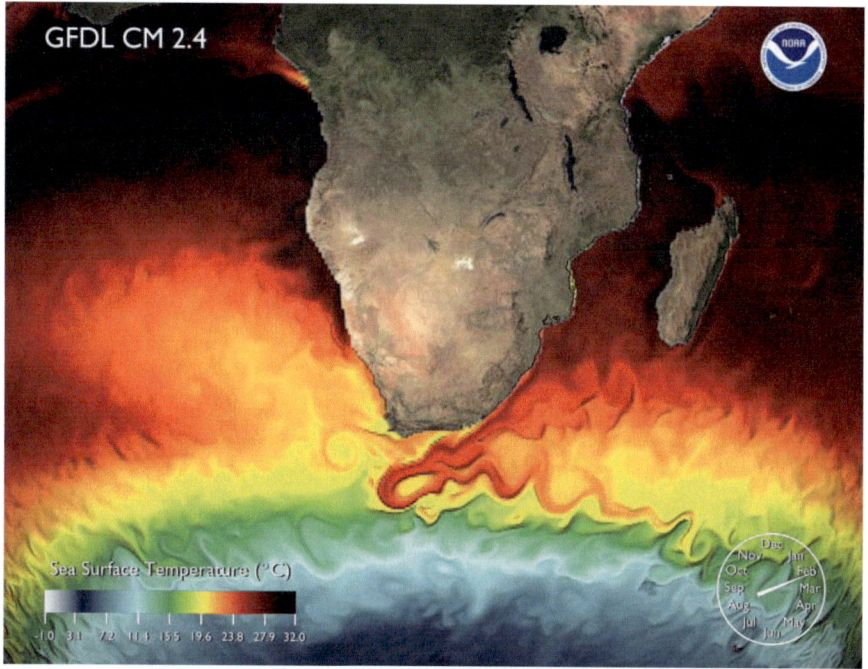

Fig. 4.2: The Agulhas current, where the Indian Ocean meets the Atlantic, produces eddies as warm and cold water mix. The eddies can combine to form larger circulations, eventually forming hurricanes, possibly as a result of negative temperature states. *Credit: NOAA*

tion of hurricanes. In the Atlantic many of these storms form as a result of eddies emerging from the Agulhas current (Fig. 4.2). Upon combining these eddies can generate powerful rotations that race across the Atlantic. Hurricanes are just one emergent phenomena into which vortex statistics can give insight.

Before we get to negative temperature states, first, we must derive the point vortex gas from first principles. The induced velocity of a three-dimensional vorticity field, ω, is given by the Biot–Savart law,

$$\mathbf{u} = -\frac{1}{4\pi} \int d^3 x' \frac{(\mathbf{x} - \mathbf{x}') \times \omega(\mathbf{x}')}{|\mathbf{x} - \mathbf{x}'|^3}. \tag{4.1}$$

Given the kernel $K = -(4\pi)^{-1}\mathbf{x}/|\mathbf{x}|^3$, this can be written as a convolution equation,

$$\mathbf{u} = K * \omega, \tag{4.2}$$

where convolution of two functions, f and g, is defined as,

$$f * g = \int d\mathbf{x}' f(\mathbf{x}') g(\mathbf{x} - \mathbf{x}'). \tag{4.3}$$

In two dimensions the kernel is the 2D vector,

$$K = \frac{1}{2\pi}(-\partial_y, \partial_x) \log|\mathbf{x}|, \quad (4.4)$$

where $(\cdot,\cdot)$ denotes a vector with two components, in this case, two derivatives both applied to the log. The vorticity is a scalar valued function ω.

Let N perfectly parallel vortices of circulations $\lambda_1, \ldots, \lambda_N$ be in an incompressible, frictionless fluid such that the vorticity field has the form,

$$\omega(x,y) = \sum_i \lambda_i \delta(x - x_i) \delta(y - y_i). \quad (4.5)$$

Evaluating (4.2) with (4.4),

$$\frac{dx_i}{dt} = \frac{1}{2\pi} \sum_{j \neq i} \frac{\lambda_j (y_j - y_i)}{r_{ij}^2} \quad (4.6)$$

and

$$\frac{dy_i}{dt} = -\frac{1}{2\pi} \sum_{j \neq i} \frac{\lambda_j (x_j - x_i)}{r_{ij}^2}, \quad (4.7)$$

are the equations of motion. These equations show how velocity in the x direction depends on the ratio of the distance in the y direction and the inverse of the square distance. For example, if two vortices have the same x coordinates but different y coordinates, then velocity can only be in the x direction. This is quite different from typical particles where like-signed repulsion tends to happen along the line connecting the two. Here repulsion pushes the vortices, not directly away from one another, but in opposite directions along the line *perpendicular* to the line connecting them.

The set of N vortices is a finite dimensional Hamiltonian system where the equations of motion can be written,

$$\begin{cases} \lambda_i \frac{dx_i}{dt} = \frac{\partial H}{\partial y_i}, \\ \lambda_i \frac{dy_i}{dt} = -\frac{\partial H}{\partial x_i}, \end{cases} \quad (4.8)$$

where H is the energy functional (4.9). Oddly, the position variables are conjugate to one another rather than position being conjugate to momentum. In fact, momentum does not appear in the equations at all. The self-energy—meaning the energy internal to the filaments rather than between filaments—of a perfectly parallel filament is generally infinite like rope held at infinite tension but "constant" in the limit as the vortex radius vanishes; hence, for an unbounded plane the energy is

$$H = -\frac{1}{2\pi} \sum_{i > j} \lambda_i \lambda_j \log r_{ij}, \quad (4.9)$$

where $r_{ij} = \sqrt{(x_i - x_j)^2 + (y_i - y_j)^2}$ is the distance between vortex i and vortex j. If we have different boundary conditions (such as a sphere [44, 132]), we replace

4.2 Negative Temperature States

$\log r_{ij}$ with a different function related to the kernel for the Biot–Savart equation under those conditions.

4.2 Negative Temperature States

Now that we understand the point vortex gas. It is time to look at negative temperature states. We are all familiar with positive temperatures, but what are negative temperatures? By negative temperatures I do not refer to temperature measures that have a negative absolute like Celcius and Fahrenheit but to measures that have a zero absolute like Kelvin. What does it mean for a system to have a negative Kelvin temperature? Since absolute zero indicates no molecular motion in the system, presumably one cannot have less than zero kinetic energy and hence less than zero motion. That definition of temperature, however, is not what is indicated here. Temperature, instead, has another definition in terms of how entropy changes with energy,

$$\frac{1}{T} = \frac{\partial S}{\partial E}, \tag{4.10}$$

which is not based on the kinetic energy at all. With this definition, the possibility of negative temperatures becomes obvious. Many systems can be devised where the components have many configurations at a low energy and fewer configurations at a higher energy. Since entropy is simply the logarithm of the number of configurations, if that number decreases with increasing energy, then of course the temperature is negative.

One of the most interesting results from this model is the discovery of negative temperatures. Negative absolute temperature tends to appear in special cases where there is an upper bound on energy, such as nuclear spin systems [125]. Therefore, although seemingly paradoxical, it is simply uncommon to everyday experience where kinetic energies of molecular matter allow for no upper bound. Negative temperatures in a 2-D Coulomb system indicate that higher energy states are favored over lower. Since the Coulomb interaction is repulsive, the highest energy states are when the vortices cluster together. And indeed, negative temperature states contain clusters. Another feature of negative temperature states is that the direction of "increasing" temperature. Negative temperatures are hotter than infinite temperature (because their energy states are higher). Thus, increasing temperature runs like so: $0 < \cdots < +T < \cdots < +\infty < -\infty < \cdots -T < \cdots -0$ where $T > 0$.

Negative temperature states have been observed in nuclear spin systems [123, 125]. In this case, we have a collection of N atoms each with two possible states spin-up $s = 1$ and spin-down $s = -1$. Given a random collection of atoms, roughly half should be spin-up and half spin-down because this is the highest entropy state. Entropy, S, decreases away from this 50/50 split with the lowest entropy states being all spin-up or all spin-down. We can induce a magnetic field, however, that encourages more atoms to be in the spin-down state. The total energy, E, is proportional to the number of atoms in the spin-up state. If we then begin adding energy by, e.g., bombarding the atoms with radiation, we can induce more atoms to flip their spins to spin-up. As the ratio approaches 50/50, the entropy increases, and, since energy and entropy are both increasing, temperature, T, is increasing as well by the thermodynamic definition $1/T = \partial S/\partial E$ (in discrete systems $1/T = \Delta S/\Delta E$). Once the ratio passes the 50/50 ratio, however, entropy begins falling while energy continues to increase. This corresponds to a negative temperature state.

Negative temperature states have several interesting properties. For example, when the temperature is negative, the energy wants to go higher, not lower. See what we mean: given the Boltzmann distribution's likelihood, $\lambda = \exp[-E/(k_B T)]$, if $T < 0$, for two energies, $E_1 > E_2$, E_1 has higher likelihood. But that's not all: negative temperatures are *hotter* than positive temperatures, where hotness indicates the direction of energy flow, assuming that heat flows from hotter to cooler systems (when specific heat is positive). In fact, one can show that $T = -\infty$ is equivalent to $T = \infty$ in terms of heat flow (essentially at infinite or minus infinite temperature all energy states are equally likely) but finite negative temperatures are *above* infinite temperature. To prove it take two systems each in equilibrium with energies E_1 and E_2, entropies S_1 and S_2. Combine the two systems. Initially, we have total energy and entropy,

$$E = E_1 + E_2 \tag{4.11}$$
$$S = S_1 + S_2. \tag{4.12}$$

The energy stays fixed while, from basic thermodynamics, the entropy is expected to increase as the combined system reaches equilibrium. Take the time derivatives,

$$\frac{dE}{dt} = \frac{dE_1}{dt} + \frac{dE_2}{dt} = 0, \tag{4.13}$$
$$\frac{dS}{dt} = \frac{dS_1}{dt} + \frac{dS_2}{dt} > 0 \tag{4.14}$$

Rewriting (4.14) in terms of temperature we have

$$\frac{dS}{dt} = \frac{dS_1}{dE_1}\frac{dE_1}{dt} + \frac{dS_2}{dE_2}\frac{dE_2}{dt} \tag{4.15}$$
$$= \left(\frac{dS_1}{dE_1} - \frac{dS_2}{dE_2}\right)\frac{dE_1}{dt} \tag{4.16}$$
$$= \left(\frac{1}{T_1} - \frac{1}{T_2}\right)\frac{dE_1}{dt} > 0 \tag{4.17}$$

where the middle equation comes from (4.13) which implies that $dE_1/dt = -dE_2/dt$.

Suppose that the second system is hotter than the first, $T_2 > T_1$, then, by (4.17), $dE_1/dt > 0$. The energy is moving from the hotter system, 2, to the cooler as expected. If, on the other hand, $T_2 < 0$ (negative) and $T_1 > 0$ (positive), then we *also* have $dE_1/dt > 0$. Energy is moving from the negative temperature system to the positive! This proves that negative temperatures are hotter than positive.

4.2 Negative Temperature States

Negative temperatures always occur when there is a positive temperature upper bound on energy. In other words, every energy, E, has a corresponding temperature, T, but systems can sometimes "run out of" positive temperatures when they hit a particular energy upper bound. The remaining energies above that threshold, $E > E_{\max}$, that do not have a positive temperature must have a negative temperature associated with them. (One can even have complex temperatures although the authors could not find reference to any physical interpretation for these in the literature.) Every familiar molecular system, of course, has unbounded energy because any molecule can, theoretically, approach the speed of light where it has infinite energy. Hence, molecules have no negative temperature states. Other systems such as the point vortex gas, however, have no concept of kinetic energy. Thus, they can have bounded energies at positive temperatures.

Now that we understand negative temperatures. Let us combine what we know with the 2D point vortex gas. Onsager employed a trick to show that 2D point vortex gas has negative temperature states. Enclose the ensemble in a box with area A. Consider the total volume of the phase-space (space over all conjugate variables),

$$\int dx_1 dy_1 \ldots dx_n dy_n = \int d\Omega = A^n. \tag{4.18}$$

Define the portion of the phase-volume where the energy is less than a value E:

$$\Phi(E) = \int_{H<E} d\Omega = \int_{-\infty}^{E} dE'\, \Phi'(E'), \tag{4.19}$$

such that $\Phi(-\infty) = 0$ and $\Phi(\infty) = A^n$.

Since the function $\Phi(E)$ is monotonically increasing, $\Phi'(E)$ is positive for all E. Also, because $\Phi(\infty) < \infty$, $\Phi'(E)$ hits its maximum at some energy $E_{\max}$. Thus, $\Phi''(E_{\max}) = 0$ and is positive below the maximum and negative above it. (This is just basic extrema calculus.) If you take the derivative of (4.19), you can see that the result, Φ', is precisely the size of the configuration space at a fixed energy E (microcanonical ensemble). Statistical mechanics tells us that the entropy is $S(E) = \log \Phi'$, so $\Phi' = \exp[S(E)]$. Temperature is defined as

$$\frac{1}{T} = S'(E), \tag{4.20}$$

and, taking the derivative once more, $\Phi''(E) = S'(E)\exp[S(E)]$, so the temperature is given by,

$$T = \frac{\Phi'}{\Phi''} = 1/S'(E). \tag{4.21}$$

We know that $\Phi'' > 0$ below the energy bound, $E < E_{\max}$, but $\Phi'' < 0$ above it, $E > E_{\max}$. That gives us the astounding conclusion, however, that when $E > E_{\max}$, $T < 0$. Hence, for energies *increasing* beyond that threshold, entropy begins to *decrease*. The temperature is negative for high energy states! Because the temperature is negative, vortices of the same sign will cluster together, possibly merging [117].

The reader might need to read through that argument a few times to absorb it completely. In fact, it left many contemporary scientists scratching their heads and wondering where the error was. It does, in fact, have several problems with it as we will get to below, but it is fundamentally true. Vortex systems do have an energy upper bound for positive temperatures when confined (as a result of that confinement in fact), and they do have negative temperature states above that energy.

Onsager's result ties into a well-known result from Kolmogorov's scale theory (discovered in the early 1940s about the same time Onsager was doing his work on the point vortex gas). Kolmogorov's scale theory is one of the foundations of turbulence theory and agrees with Onsager's results. It goes as follows: In three-dimensional systems, energy of turbulent flow dissipates to smaller and smaller length scales, called the *Kolmogorov energy cascade*, until it reaches the molecular and dissipates as heat. Thus, Kolmogorov's energy cascade gives the intuitive result we have from positive temperature systems—things tend to dissipate their energy at small scales. They slow down because of friction and heat loss.

Kolmogorov's theory from 1941 gives a formula for the cascade,

$$E(k) = C\varepsilon^{2/3}k^{5/3}, \tag{4.22}$$

for the energy spectra [80]. Here C is a constant and ε is the mean dissipation per unit mass. The details of how this formula is arrived at are not important here, as there are many good books on turbulence that discuss it. Looking at the formula, however, it is easy to see why the energy cascades to lower length scales. Because the exponent on the wavenumber is positive, energy increases at higher and higher wavenumbers, which imply lower and lower wavelengths, hence smaller scales. Thus, at the molecular level, where frequencies are greatest, one expects the greatest energy.

In two-dimensional systems, however, our intuition is betrayed. Energy cascades in the opposite direction, called the *inverse* energy cascade, given by the formula,

$$E(k) = C'\varepsilon^{2/3}k^{-5/3}. \tag{4.23}$$

Because the exponent on the wavenumber is now negative, smaller wavenumbers give larger energies. Instead of tearing down structures, two-dimensional turbulence builds them up to higher and higher length scales. Intuitively, it is easy to see that the entropy is decreasing with increasing energy as well, since many small structures have more degrees of freedom than a few large ones. This explains, for example, why small collections of eddies forming off the South African coast can combine to form massive Atlantic hurricanes. Ultimately, of course, these are three-dimensional structures and, given enough time, their energy cascade will reverse, but because the third dimension is "squashed" compared to the other two, it takes a lot longer.

While ultimately vindicated, Onsager's analysis in the 1940s was not only mathematically non-rigorous but made dubious physical assumptions. First, because they move and represent the entire flow discretely, point vortices are compressible while the underlying Euler fluid is incompressible [118]. Second, the negative temperature analysis is based on the confining volume which, while admissible in the lab, hardly describes meterological phenomena which, when relatively small, may be considered unbounded, and when large (such as hurricanes) may be considered on a rotating sphere [44]. Onsager was, in fact, careful not to make use of the actual Hamiltonian form he gives in his seminal paper, which is for the unbounded plane. By explaining that the Hamiltonian derives from Laplace's equation and that a suitable alternative Green's function solution may be chosen to replace it, he implies that his arguments are valid for bounded domains as well. This does not, however, place the arguments on much sounder footing, since he assumes a boundary exists. Also, at the time, turbulence was thought to be far from equilibrium, making the assumption of equilibrium statistical ensembles inapplicable. Later this belief was shown to be false as time scale arguments were introduced—indeed, the universe itself is not at equilibrium (entropy is not yet maximal since the universe has not yet suffered heat death), hence, as we discussed in the previous chapter, *any* equilibrium statistical mechanics has a hidden time scale assumption. That turbulence has a short time scale makes no difference as long as the structures considered form on a much shorter time scale than the overall dissipation.

4.3 The Guiding Center Model

It took over two decades to put Onsager's insights onto a slightly more solid footing. It is possible that the renewed interest was related to the rising concern with alternative power sources as the 60s became the 70s. Perhaps it is more than a coincidence, indeed, that two of the most important papers on this topic, [46, 70], were released at approximately the same time as the 1973 oil embargo, coinciding closely with the startling achievement of significant electron temperatures in Tokamak magnetic nuclear fusion devices. For the first time it seemed possible to generate power from deuterium fuel, but there were stumbling blocks, and these had to do directly with turbulence in neutral and electron plasmas.

The study of plasmas and fluids have in recent years become nearly synonymous in the theoretical fluids community. On their face, they seem very different however. Plasmas occur at very high temperatures often in the range of 10–100 MK, the latter being the target temperature for magnetic nuclear fusion [151]. At these temperatures electrons are released from atomic nuclei and roam free at high speeds. Therefore, one major difference between plasmas and fluids is that plasmas are charged fluids. A neutral plasma consists of two charged fluids which are separate but mixed together. A neutral fluid, on the other hand, is a single fluid containing molecules where electrons are firmly bonded to their nuclei. Because plasmas are charged, Maxwell's equations become involved, and, indeed, without significant simplification the behavior of plasmas seems completely different from that of ordinary fluids.

The guiding center model is such a simplification. The guiding center model is essentially a parallel vortex model for a neutral, two-component plasma. Given long filaments of charge aligned parallel to a uniform magnetic field **B** and moving under their mutual electric field **E**, the equations of motion for the filaments are

$$e_i \frac{dx_i}{dt} = \frac{1}{B}\frac{\partial H}{\partial y_i}, \quad e_i \frac{dy_i}{dt} = -\frac{1}{B}\frac{\partial H}{\partial x_i}, \tag{4.24}$$

and

$$H = \sum_{i<j} -\frac{2e_i e_j}{l} \log|\mathbf{r}_i - \mathbf{r}_j|, \tag{4.25}$$

where $B = |\mathbf{B}|$, l is the length and e_i is the charge [46]. These equations are identical to the ones above for parallel vortex filaments in an ordinary fluid up to a constant factor. These long filaments are columns of electrons and deuteron ions. ("Column" is yet another word we will use for filaments but only in the context of charged fluids.) The particles orbit the guiding center of the columns at a distance known as the Larmor radius, a, at a frequency $\omega = \sqrt{ne^2/(m\varepsilon_0)}$, where n is the number density, e is the electron charge, m is the particle mass, and ε_0 is the permittivity of free space. Hence the electrons behave in much the same way as the molecules of fluid in the ordinary fluid vorticity model of Onsager. The main difference is that it is the charge and not the vorticity that represents the strength of each column.

Edwards and Taylor [46] go through a nice, exact, constructive calculation for an equation of state from the microcanonical ensemble, which we will give here. For N columns at energy E in a confining volume V, let the phase-space density be

$$\Omega(E,V,N) = \frac{d\Phi}{dE} = \int d\Omega \, \delta(E - H), \tag{4.26}$$

such that the entropy is

$$S(E,V,N) = \log \Omega. \tag{4.27}$$

Taking the Fourier transform of the delta function integrand,

$$\Omega = \int \frac{d\lambda}{2\pi} dx_1 dy_1 \cdots dx_N dy_N \exp[i\lambda(E + \sum_{i \neq j} e_i e_j \log r_{ij})] \tag{4.28}$$

Now, scale by the volume, $r \to r'V^{1/2}$ and the integral becomes

$$\Omega = \frac{V^{2N}}{2\pi} \int \frac{d\lambda}{2\pi} dx_1 dy_1 \cdots dx_N dy_N \exp[i\lambda(E + \sum_{i \neq j} \frac{e_i e_j}{2} \log V + \sum_{i \neq j} e_i e_j \log r_{ij})]. \tag{4.29}$$

If $e_i = \pm e$, since we have an equal number of positive and negative columns $\sum_{i,j} e_i e_j = 0$, hence, $\sum_{i \neq j} e_i e_j = -Ne^2$. The phase-space density becomes

$$\Omega = \frac{V^{2N}}{2\pi} \int \frac{d\lambda}{2\pi} \exp\left(i\lambda E - \frac{i\lambda Ne^2}{2}\log V\right)$$
$$\times \int dx_1 dy_1 \ldots dx_N dy_N \exp\left(i\lambda \sum_{i \neq j} \log r_{ij}\right). \tag{4.30}$$

4.3 The Guiding Center Model

Given the formulas for pressure,

$$P = -\left(\frac{\partial E}{\partial V}\right)_{S,N} = \frac{\partial S}{\partial V}\left(\frac{\partial S}{\partial E}\right)^{-1}, \quad (4.31)$$

and temperature,

$$\frac{1}{T} = \frac{\partial S}{\partial E}, \quad (4.32)$$

we can take the derivative of (4.30) with respect to V and E to arrive at the equation of state,

$$PV = 2NT(1 - e^2/2T) \quad (4.33)$$

which is sufficient to show negative pressure but not negative temperature. The regime where the pressure is negative, however, is unattainable. The regime where temperatures are negative is attainable though. To show negative temperature states explicitly requires taking an approximation for large N. The basic outline is to take the Fourier series of the density of positive and negative charges, respectively:

$$p_k = \frac{e}{V}\sum_{+}\exp[i\mathbf{k}\cdot\mathbf{r}_i], \quad q_k = \frac{e}{V}\sum_{-}\exp[i\mathbf{k}\cdot\mathbf{r}_j],$$

then let

$$\rho_k = p_k - q_k, \quad \eta_k = p_k + q_k,$$

so that the energy becomes

$$H = 2\pi V \sum_k \left(|\rho_k|^2 - \frac{4\pi N e^2}{V^2}\right).$$

The essential step is then to change variables from $\mathbf{r}_i$ and $\mathbf{r}_j$ to ρ_k and η_k and do the integration, where for large N the Jacobian is

$$J = V^{2N}\prod \frac{V^2}{2Ne^2}\exp\left[-\frac{V^2}{2Ne^2}(|\rho_k|^2 + |\eta_k|^2)\right].$$

From these expressions, one can apply the saddle point method to evaluate (4.30). From that one obtains the entropy and, thereby, the temperature. We will leave the details as an exercise for the reader.

We mention that this analysis still suffers from many of the same problems as Onsager's analysis, such as having a boundary and that the fluid is not clearly incompressible. Hence, conservation laws are not respected. Another problem that we will show is quite important is the assumption of perfectly parallel vortices. Although this is frequently seen as a reasonable assumption, the lack of a self-energy term has a significant impact on the statistics. In later chapters, we will investigate quasi-2D filaments (with some variation in the third dimension) and show some startling results.

4.4 Continuous Vorticity

Forty years after his initial insight, Onsager was vindicated with relative mathematical rigor with the Robert–Miller model [106, 130]. Here we will present an abbreviated version of Miller's rigorous mean-field theory (essentially a lattice renormalization) of vorticity in an Euler fluid. Miller's approach is to present the ensemble, not as discrete vortices that move but as a vorticity field on a lattice. Suppose we have a disk of radius 1, D, and we impose two-dimensional inviscid flow on this disk with velocity field **u**. The Hamiltonian is given by

$$\mathcal{H} = \frac{1}{2} \int_D d^2 x |\mathbf{u}|^2. \tag{4.34}$$

With the equation,

$$-\nabla^2 \psi = \omega, \tag{4.35}$$

where $\mathbf{u} = \nabla \times \psi$ and ψ is the stream function,

$$\mathcal{H} = \frac{1}{2} \int_D d^2 x [\psi \cdot \omega - \nabla \cdot (\mathbf{u} \times \psi)]. \tag{4.36}$$

The second term in the integrand becomes a surface term on the boundary, $B = \partial D$, which we assume constant.

$$\mathcal{H} = \frac{1}{2} \int_D d^2 x \psi \cdot \omega. \tag{4.37}$$

Using the Green's function for the Laplacian,

$$\mathcal{H} = -\frac{1}{4\pi} \int_D d^2 r \int_D d^2 r' \, \omega(\mathbf{x}) \omega(\mathbf{x}') \log |\mathbf{r} - \mathbf{r}'|, \tag{4.38}$$

up to an additive constant.

On a lattice regularization with spacing a this becomes

$$\mathcal{H}_a = -\frac{a^4}{4\pi} \sum_{i \neq j} \omega_i \omega_j \log |\mathbf{r}_i - \mathbf{r}_j| + (\text{self-energy}), \tag{4.39}$$

where i and j count over the lattice points and $\omega = |\omega|$. The self-energy, we can show, vanishes as $a \to 0$. This Hamiltonian resembles the one for a vortex gas but it has an underlying lattice which allows conservation laws such as incompressibility to be maintained and does away with problems with using delta functions to discretize the vorticity in the Onsager model.

Rather than go into a detailed calculation, it suffices to give an intuitive feel for Miller's argument for the validity of the Onsager partition function in the thermodynamic limit. Miller's prescription is to define a vorticity distribution function, $G(\omega)$, as the measure (area) of vorticity less than ω on the unit disk. This function is preserved over time evolution. The vorticity can change locations but

4.4 Continuous Vorticity

the amount less than ω cannot cover less area because of incompressibility. The vorticity is also bounded $\omega < |\omega|_{max}$. The vortex density is given by the derivative, $g(\omega) = G'(\omega)$. (Another way to look at it is that if we take the vortex density g, then $G(\omega) = \int_{\omega_{min}}^{\omega} d\omega' g(\omega')$ is the cumulative density function.)

As an example, consider the disk with a density function,

$$g(\omega) = (\pi - \alpha)\delta(\omega) + \alpha\delta(\omega - 1). \tag{4.40}$$

Since the area of $\Omega = \int d\omega' g(\omega') = G(\omega_{max})$ is π, this density function assigns a density of $\omega = 1$ to an area α and causes the vorticity to vanish on the rest of the disk. Like energy, vorticity density is a macrostate in that it does not specify precisely where the density is applied. Given that there are $N = \pi/a^2$ lattice points, the number of configurations is the number of ways that 1's can be assigned to α/a^2 points and 0's assigned to $(\pi - \alpha)/a^2$ points. Note that this combinatorial problem is the reverse of the point vortex problem. Whereas with point vortices we are assigning points of fixed vorticity to a continuous plane, on the lattice we are assigning different vorticities to points that are at fixed locations.

We now need to show that the statistical equilibrium of 2D Euler flow is well defined. The crux of the proof that, for a given vortex density, g, the partition function,

$$\int D^g \omega \exp[-\mathcal{H}(\omega)/T], \tag{4.41}$$

is well defined as the lattice spacing, $a \to 0$, and, hence, the statistical mechanics of 2D Euler flow is well defined comes from an exact mean-field theory. An exact mean-field theory shows that the equilibrium states of a system of interacting particles or lattice points is the same as the equilibrium states of these individual particles or lattice points independently interacting with an external field. This theory requires removing correlations between points in the disk which represent the interaction between particles.

In order to remove correlations and carry out the evaluation of the configuration space, take the lattice with spacing a and impose a length scale l (Fig. 4.3). State that within this length scale lattice points are independent of one another. This means that the disk's lattice has been divided into sub-lattices such that $l \gg a$, so that each box of side l has l^2/a^2 lattice points in it. Now given a distribution function $G(\omega)$, show that there exists an l and $|\omega|_{max}$ such that the $|\mathcal{H} - \mathcal{H}_l| < \varepsilon$, where $\mathcal{H}_l$ is our energy when we remove the interactions within the sub-lattices. The lattice spacing is still a, so this is a coarser grained energy functional than $\mathcal{H}_a$. As $l \to 0$, we can show that this coarse grained energy approaches the continuum energy, i.e., $\mathcal{H}_l \to \mathcal{H}$, uniformly. In other words, ε does not depend on $G(\omega)$. Because of uniform convergence, the actual vorticity field is irrelevant to the convergence, hence correlations on scales less than l are irrelevant as well because they vanish in the limit. The entropy can then be calculated as if the system within each box of side l were an ideal gas.

This is the beauty of the mean-field theory, we get the equilibrium statistics we want while sweeping all the badly behaved small scale behavior under the rug.

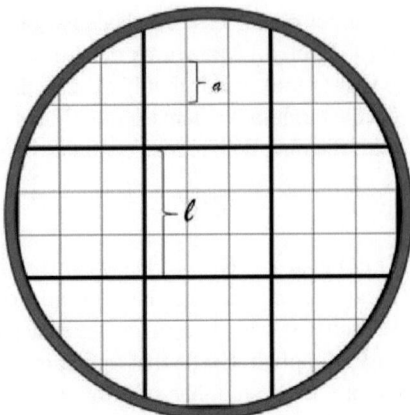

Fig. 4.3: The two length scales define different behaviors. Within the boxes of side l, we have non-interacting ideal gases of vorticity on the lattice points with vortex separation a. Between the boxes we have Coulomb interactions

We can first take $a \to 0$ without worrying about interactions, then take $l \to 0$. The reason for the two discretizations is that, with the smallest discretization we want to average out the short length scale behavior of vortices. At these scales vortices behave like hard-core particles in an ideal gas. At longer scales, Coulomb interactions dominate. The separation of length scales is critical to the exactness of the mean-field theory. A complete derivation can be found in [107], Sect. V.

The essence of the argument is that statistics of 2D Euler flows is a consequence of long-wave, large-scale, coarse grained behavior. The potential for blow-ups and overwhelming interactions at small scales is eliminated by the regularization and mean-field theory. Negative temperature states also can be shown to exist with continuous and finite vorticity functions [106]. These negative temperature states now arise honestly from the 2D Euler flow equations, however, and show that Onsager's insight was, in fact, correct, although it took 50 years to show rigorously.

Chapter 5
Curved Filaments

Now that we have looked at straight, parallel filaments, it is time to look at curved ones. Two-dimensional vorticity is a very special case of fluid flow that tends to apply to thin films, atmospheres, and structures with high angular momentum. In aeronautical or hydrodynamical applications such as turbulent sheer flow, vortices can be filaments or rings with a more complex topology such as trefoil knots (Fig. 5.1).

The study of non-2D vortex filaments is relatively young, however, because of all the additional complexities that non-2D models introduce. Unlike point vortices, vortices that are non-2D no longer behave like charged particles with a simple interaction potential. As we shall see, however, they are essential to understanding the behavior of many types of flows that are not two-dimensional.

5.1 Motion of a Vortex Filament

We approach the motion of curved vortex filaments using asymptotic matching. Asymptotic matching is a kind of boundary layer method where a complicated internal flow model yields to a simpler external flow model, and we want to abstract away the internal model via asymptotic analysis, i.e., when the size of the internal region goes to zero and the vortex becomes a line, what is the model for the behavior of that region in the wider world? Of course, there is no such thing as a zero volume vortex filament in the real world, so we are really interested in the case where the width of the filaments gets very, very small. For our introduction to this matching procedure we look at only a single filament. In the next section, we will generalize this to many interacting filaments and obtain a statistical ensemble.

Suppose we have an inviscid, irrotational fluid containing a single vortex filament. Inside the filament we have a viscous, rotating fluid. We justify the inviscid assumption outside the core because viscous forces are assumed to be negligible there compared to inertial forces. Inside the filament, however, we have a large radial

58   5 Curved Filaments

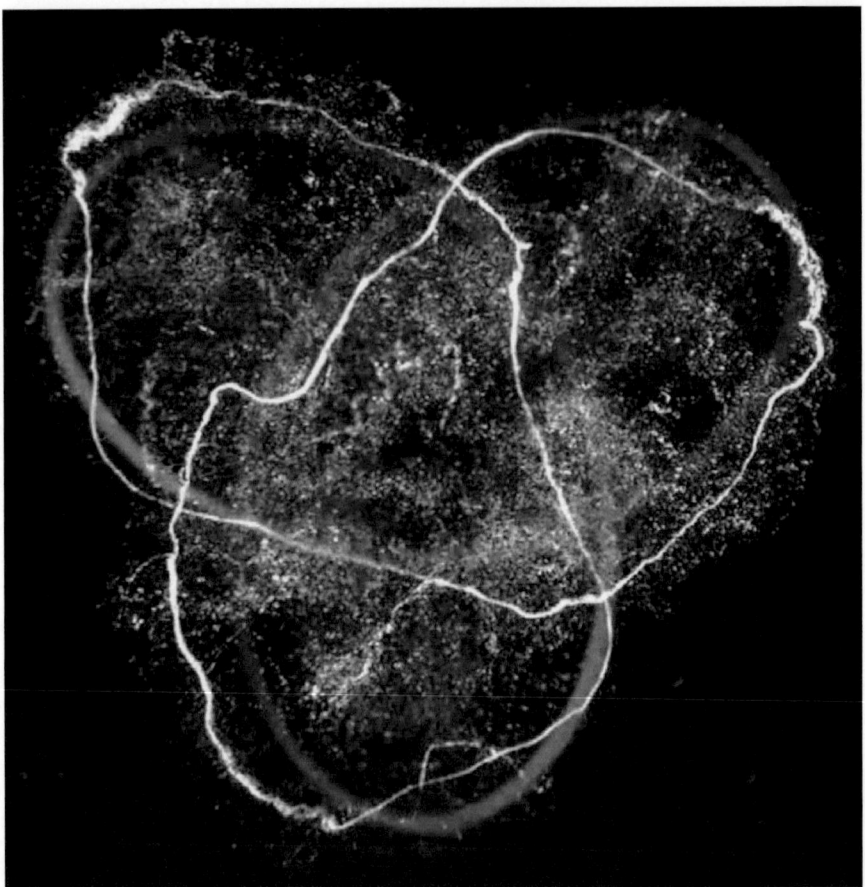

Fig. 5.1: Recently liquid vortices have been formed into trefoil knots, an experimental achievement of a theoretical vision dating back 150 years. *Credit: Kleckner and Irvine. Reprinted by permission from Macmillan Publishers Ltd:* Nature Physics *[76], copyright 2013*

pressure gradient pushing fluid particles toward the center of the filament which creates the centripetal force necessary to maintain the vortex. Here viscous forces are no longer negligible and the full Navier–Stokes equations must be applied.

Because we have these two distinct regions (1) the narrow vortex core and (2) the inviscid fluid, we can calculate the motion of the vortex filament using a matching procedure where motion due to interior forces comes from the Navier–Stokes equations and motion due to exterior forces comes from the Biot–Savart equation. Let the vortex filament or line be the center line of the vortex core. Suppose that the vortex filament's 3D spatial position is given by the curve with parameter s moving in time t, $\mathbf{X}(s,t)$. Let the velocity of the fluid outside the vortex filament be $\mathbf{Q}$. Split the velocity into two parts (1) the velocity that the vortex filament induces, $\mathbf{Q}_1$ and

5.1 Motion of a Vortex Filament

(2) the background flow field (everything else) Q_2. As with parallel filaments, the velocity that the filament induces on the flow at a point $\mathbf{P}$ is given by the Biot–Savart equation. The Biot–Savart law comes from the equations for an incompressible, inviscid fluid. Since, from incompressibility, $\nabla \cdot \mathbf{Q}_1 = 0$, the velocity can be written in terms of a stream vector function, ψ, $\mathbf{Q}_1 = \nabla \times \psi$. The stream vector is the "vector potential" of the velocity field. Using the definition of vorticity, $\omega = \nabla \times \mathbf{Q}_1$, we can describe the vorticity by the Poisson equation,

$$\omega = \nabla \times \mathbf{Q}_1 = -\nabla^2 \psi. \tag{5.1}$$

The Biot–Savart law (5.2) is the solution to this Poisson's equation:

$$\mathbf{Q}_1(\mathbf{P},t) = -\frac{\Gamma}{4\pi} \int \frac{[\mathbf{P} - \mathbf{X}(s,t)] \times ds'}{[\mathbf{P} - \mathbf{X}(s',t)]^3}, \tag{5.2}$$

where $\Gamma > 0$ is the circulation,

$$\Gamma = \oint_C \mathbf{Q}_1 \cdot d\mathbf{l}, \tag{5.3}$$

for a closed curve about the filament, C. The Biot–Savart equation gives the velocity that a vortex filament induces on the fluid external to itself, including the velocity one part of the filament induces on another part of the filament.

Every asymptotic expansion has a small parameter. In this case, the small parameter for the expansion is $\varepsilon = (\nu/\Gamma)^{1/2}$, the square root of the ratio of the viscosity to the circulation (inverse Reynolds number). Therefore, the solution is valid for cases where the kinematic viscosity in the core is much smaller than the vortex circulation, $\nu \ll \Gamma$. Note that both kinematic viscosity and circulation have units of area over time so the parameter is non-dimensional. This parameter essentially measures the dissipation of the vortex. By Helmholtz's laws a vortex filament persists indefinitely in the absence of viscosity (just as an object will rotate indefinitely in the absence of friction by conservation of angular momentum). The higher the viscosity the greater force there is for dissipation. Counteracting this viscous dissipation is the speed at which the vortex rotates. Slower vortices dissipate more quickly because they have less radial pressure gradient keeping them intact; hence less energy needs to dissipate into viscosity to achieve the dispersion of the filament. Assuming this is a small parameter suggests that the filament is relatively persistent. As $\varepsilon \to 0$, either the viscosity goes to zero or the circulation goes to infinity (or both). This is equivalent to taking the size of the vortex core to zero.

5.1.1 The Curvilinear Formulation

Since we are dealing with only one vortex filament, it helps to use a coordinate system that has the filament itself as an axis (Fig. 5.2). (This approach is due to Callegari and Ting [29].) Given the curve $\mathbf{X}(s,t)$, the curvilinear coordinate system is fixed to the curve running through the center of the vortex and moves with it. The unit normal, $\hat{\mathbf{n}}$, binormal, $\hat{\mathbf{b}}$, and tangential, $\hat{\boldsymbol{\tau}}$, coordinate vectors and the linear strain, σ, curvature, k, and torsion, T, are given by the usual Serret–Frenet formulas. Since this book emphasizes explanation over computation, we will not give them here although any text on vector calculus will have them for a curve $C(s)$. The three unit basis vectors, as well as strain, curvature, and torsion are all functions of time as well as the curve parameter since the filament is not stationary.

We define the radial, circumferential, and tangential coordinates (r, θ, s) and the unit basis vectors $\hat{\mathbf{r}}, \hat{\boldsymbol{\theta}}, \hat{\boldsymbol{\tau}}$ of an orthogonal coordinate system also tied to the reference curve. Let $\phi = \theta + \theta_0$ be the angle between the normal vector, $\hat{\mathbf{n}}$, and the radial vector, $\hat{\mathbf{r}}$, and $\theta_0 = -\int ds\, \sigma T$ where T is the torsion of the reference curve. Then we have

$$\hat{\mathbf{r}} = \hat{\mathbf{n}}\cos(\phi) + \hat{\mathbf{b}}\sin(\phi) \tag{5.4}$$

$$\hat{\boldsymbol{\theta}} = \hat{\mathbf{b}}\cos(\phi) - \hat{\mathbf{n}}\sin(\phi). \tag{5.5}$$

The scale factors are

$$h_1 = 1, \quad h_2 = r, \quad h_3 = \sigma[1 - kr\cos(\theta + \theta_0)], \tag{5.6}$$

where $k(s,t)$ is the curvature and $\sigma = |\mathbf{X}_s(s,t)|$ is the linear strain. (This is all standard curvilinear coordinate calculus.) We apply the coordinate system (r, θ, s) instead of (r, ϕ, s), even though the latter would give simpler definitions, because it is not orthogonal (necessarily). If we define the azimuthal coordinate as $\theta = \phi - \theta_0$, then the coordinate system becomes orthogonal.

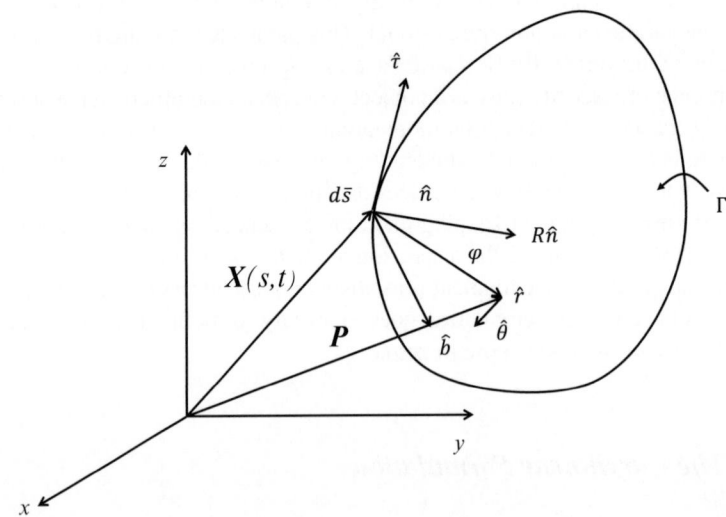

Fig. 5.2: The curvilinear coordinate system simplifies the matched asymptotic expansion [29]

5.1 Motion of a Vortex Filament

This special coordinate system is ideal for studying the motion of the filament because a position coordinate, **P**, is easily expressed as

$$\mathbf{P} = \mathbf{X}(s,t) + r\hat{\mathbf{r}}(\theta,s,t), \tag{5.7}$$

where s is chosen so that r is minimal, i.e., the point **X** is the closest point on the vortex line to **P**. It is also necessary for the matching procedure.

Every boundary layer problem has an inner and outer solution. The "outer" solution comes from the Biot–Savart equation (5.2) and gives the following velocity as $r \to 0$ [126]:

$$\mathbf{Q}_1(P,t) = \frac{\Gamma}{2\pi r}\hat{\boldsymbol{\theta}} + \frac{\Gamma}{4\pi R(s,t)}\left[\ln\left(\frac{R}{r}\right)\right]\hat{\mathbf{b}} + \frac{\Gamma}{4\pi R}(\cos\phi)\hat{\boldsymbol{\theta}} + \mathbf{Q}_f, \tag{5.8}$$

where $R(s,t)$ is the local radius of curvature and $\mathbf{Q}_f$ is the part of $\mathbf{Q}_1(P,t)$ that is finite as $r \to 0$. The velocity on the vortex line itself is undefined, so, as expected, a thin boundary layer between the inviscid fluid and the line is necessary to keep things from blowing up.

Inside the boundary layer, the Navier–Stokes equations determine velocities. The velocity at any point in the flow **q** is given by

$$\mathbf{q} = \frac{d\mathbf{X}}{dt} + \mathbf{V}, \tag{5.9}$$

where

$$\mathbf{V} = u\hat{\mathbf{r}} + v\hat{\boldsymbol{\theta}} + w\hat{\boldsymbol{\tau}}, \tag{5.10}$$

is the flow velocity relative to the filament.

The Navier–Stokes equation in this coordinate system is,

$$\frac{d^2\mathbf{X}}{dt^2} + \left(w - \frac{r}{h_3}\hat{\mathbf{r}}_t \cdot \hat{\boldsymbol{\tau}}\right)\frac{d^2\mathbf{X}}{dt\,ds} + \frac{d\mathbf{V}}{dt} = -\nabla P + \frac{v}{h_3}\frac{d}{ds}\left(\frac{1}{h_3}\frac{d^2\mathbf{X}}{dt\,ds}\right) + v\Delta\mathbf{V}, \tag{5.11}$$

where the fluid density has been normalized and P is pressure. The ∇ and Δ operations are with respect to the curvilinear coordinate system and can be found using the scale factors with standard methods for generalized coordinate systems. The continuity equation is

$$r\left(\frac{dw}{ds} + \frac{d^2\mathbf{X}}{dt\,ds}\right) + \frac{d}{dr}(ruh_3) + \frac{d}{d\theta}(h_3 v) = 0. \tag{5.12}$$

For details on how these are derived, see [29].

By Helmholtz's laws of vortex motion, fluid particles that are on the filament are required to stay on the filament. Hence,

$$\frac{d\mathbf{X}}{dt} \cdot \hat{\mathbf{n}} = \mathbf{q} \cdot \hat{\mathbf{n}}, \quad \frac{d\mathbf{X}}{dt} \cdot \hat{\mathbf{b}} = \mathbf{q} \cdot \hat{\mathbf{b}}, \quad \frac{d\mathbf{X}}{dt} \cdot \hat{\boldsymbol{\tau}} = 0. \tag{5.13}$$

In other words, using (5.9) and the above equation, $u, v = 0$, which means that fluid particles on the filament can only move tangentially to the filament and never leave it.

5.1.2 A Small Example of Matched Expansion

For those who have never seen asymptotic matching for boundary layer problems, we offer a small demonstration of the procedure. Those familiar with boundary layer matching may want to skip to the next section.

Asymptotic matching is essentially a perturbation matching procedure. Take the example,

$$\varepsilon y'' + (1+\varepsilon)y' + y = 0, y(0) = 0, y(1) = 1, 0 < \varepsilon \ll 1, \quad (5.14)$$

The "outer" solution is to take the first order approximation,

$$y' + y = 0, \quad (5.15)$$

which has the solution

$$y = Ce^{-t}. \quad (5.16)$$

Because we changed the order of the equation, though the boundary values are over specified. In order to satisfy them we need $C = 0$ for $y(0) = Ce^{-0} = C = 0$ and $C = e$ for $y(1) = Ce^{-1} = C/e = 1$. This means that there must be a boundary layer at one of the endpoints, i.e., a small interval that allows for a transition from one boundary condition to another. Suppose this layer is at the $t = 0$ side, and we keep the boundary solution such that $C = e$. Therefore, the outer solution is

$$y_{\text{outer}} = e^{1-t}. \quad (5.17)$$

Rescale the independent variable as $\tau = t/\varepsilon$. The original problem is now,

$$y''(\tau) + \varepsilon^{-1}(1+\varepsilon)y'(\tau) + y(\tau) = 0. \quad (5.18)$$

Taking the leading order terms we now have

$$y'' + y' = 0, \quad (5.19)$$

which solves to $y = A - Be^{-\tau}$. Since $y(0) = 0$, $A = B$. So the "inner" solution is

$$y_{\text{inner}} = B(1 - e^{-\tau}) = B(1 - e^{-t/\varepsilon}). \quad (5.20)$$

Now we "match" the two solutions which means that we want the inner and outer solutions to agree at some point outside but near the boundary layer as $\varepsilon \to 0$. Choose $\varepsilon \ll \sigma \ll 1$ Let

5.1 Motion of a Vortex Filament

$$\lim_{\varepsilon,\sigma \to 0} y_{\text{inner}}(\sigma) = \lim_{\varepsilon,\sigma \to 0} y_{\text{outer}}(\sigma). \tag{5.21}$$

Substituting in the equations obtained already,

$$\lim_{\varepsilon,\sigma \to 0} B(1 - e^{-\sigma/\varepsilon}) = \lim_{\varepsilon,\sigma \to 0} e^{1-\sigma}. \tag{5.22}$$

Since $\sigma \gg \varepsilon$, $\sigma/\varepsilon \to 0$ and

$$B = e. \tag{5.23}$$

There are different matching methods as well as higher order methods, but one convenient method is uniformity, where we simply add the inner and outer solutions together and subtract their overlap. The final solution to the entire equation is the sum of the inner and outer solution minus the overlapping value, y_{overlap}. Inside the boundary layer, $y_{\text{outer}} \sim y_{\text{overlap}}$ while outside it we have $y_{\text{inner}} \sim y_{\text{overlap}}$. Plugging $t = 0$ which is inside the boundary layer, into y_{outer}, we get $y_{\text{outer}}(0) = e$. Plugging $t = 1$ into y_{inner}, we have $y_{\text{inner}}(1) = e(1 - e^{-1/\varepsilon}) \to e$ as $\varepsilon \to 0$. Thus, $y_{\text{overlap}} = e$ and the overall uniform asymptotic solution is

$$y(t) = y_{\text{inner}} + y_{\text{outer}} - e = e(1 - e^{-t/\varepsilon}) + e^{1-t} - e = e^{1-t} - e^{1-t/\varepsilon}. \tag{5.24}$$

Notice that this solution now meets both boundary requirements to order ε.

5.1.3 Matched Equations

Although more complex by far and involving partial differential equations rather than ordinary, the matching procedure for the vortex filament is essentially the same as the small example above. The viscous fluid of the vortex core acts as the boundary layer between the inviscid fluid and the vortex filament singularity at the center of the core. The inner solution to the Navier–Stokes equation (5.11) for the viscous core and the outer solution to the Biot–Savart (5.2) must blend into each other such that the velocity field is continuous and differentiable across the boundary layer. Assuming the vortex core is slender and the Reynolds number high, we can expand the solution to each in terms of our small parameter $\varepsilon = (v/\Gamma)^{1/2}$ and match them term by term to create a complete solution to the motion of the vortex filament.

To obtain the inner solution, we first rescale the radial curvilinear coordinate,

$$\bar{r} = r/\varepsilon. \tag{5.25}$$

This rescaling is the main reason for the curvilinear coordinate system. With this rescaled coordinate, the Navier–Stokes equation is expanded out in powers of ε and solved with mixed boundary conditions that the flow field and its derivative vanish at infinity.

This is a first order expansion. Therefore, the motion of the filament is given by a perturbation series:

$$\mathbf{X}(s,t) = \mathbf{X}^{(0)}(s,t) + \varepsilon \mathbf{X}^{(1)}(s,t) + \cdots.$$

Essentially, what one does now is derive from the Navier–Stokes equation (5.11) and the continuity equation (5.12) new equations for the inner flow which are to leading order (which turns out to be ε^{-1}) and match the terms to the equations for the outer flow to the same order [29, 144]. After a few pages of calculations (which the reader is invited to read in the cited papers), one arrives at the following PDEs in the lowest order of the curve's motion:

$$\frac{\partial \mathbf{X}^{(0)}}{\partial t} \cdot \hat{\boldsymbol{\tau}}^{(0)} = 0, \tag{5.26}$$

$$\frac{\partial \mathbf{X}^{(0)}}{\partial t} \cdot \hat{\mathbf{n}}^{(0)} = \mathbf{Q}_0 \cdot \hat{\mathbf{n}}^{(0)}, \tag{5.27}$$

$$\frac{\partial \mathbf{X}^{(0)}}{\partial t} \cdot \hat{\mathbf{b}}^{(0)} = \mathbf{Q}_0 \cdot \hat{\mathbf{b}} + \frac{\Gamma}{4\pi R^{(0)}} \ln \frac{R^{(0)}}{\varepsilon} + C_1(s,t). \tag{5.28}$$

Here $\mathbf{X}^{(0)}$ is the leading order of the line's parameterization such that the inner and outer flow are coupled by the matching procedure, $R^{(0)}$ is the leading order term of the radius of curvature of the reference line, $C_1(s,t)$ is a function representing the viscous flow in the core, and $\mathbf{Q}_0$ is the finite part of the velocity as the core size vanishes. The core function $C_1(s,t)$ is given by

$$C_1(s,t) = \frac{k^{(0)}\Gamma}{2\pi} \left[\frac{1}{2} \lim_{\tilde{r} \to \infty} \left(\frac{4\pi^2}{\Gamma^2} \int_0^{\tilde{r}} \xi (v^{(0)} d\xi - \ln \tilde{r}) - \frac{1}{4} - \frac{4\pi^2}{\Gamma^2} \int_0^\infty \xi (w^{(0)})^2 d\xi \right) \right].$$

Here we have $k(s,t) = k^{(0)} + \varepsilon k^{(1)} + \cdots$ is the curve's curvature, $v(s,t) = \varepsilon^{-1} v^{(0)} + v^{(1)} + \varepsilon v^{(2)} + \cdots$ is the $\hat{\theta}$ component of the relative fluid velocity, $\zeta^{(0)} = \frac{1}{\tilde{r}} \frac{\partial}{\partial \tilde{r}} (\hat{r} v^{(0)})$ is the lowest order axial vorticity, and $w(s,t) = \varepsilon^{-1} w^{(0)} + w^{(1)} + \varepsilon w^{(2)} + \cdots$ is the $\hat{\tau}$ component of the relative velocity (see 5.10). The initial velocity conditions determine the core structure. In statistical equilibrium, C_1 is often taken to be constant although it could also be drawn from a probability distribution.

The take-away from this section is not necessarily how to do asymptotic matching calculations which, frequently, run into pages of calculations, but to understand how we can achieve a rigorous definition of a vortex filament from first principles. Matching procedures also apply to other filament structures including electron columns in plasmas and quantum vortex filaments, although the mathematics, particularly in the core, is different. Nowadays asymptotic procedures can be done largely by symbolic manipulation software, and the impetus of the researcher is less in calculation and more in the cleverness and applicability of the asymptotic assumptions.

5.2 Nearly Parallel Vortex Filaments

Now that we have looked at one curved filament, it is time to look at an ensemble of filaments. In order to do that, we are going to make some strict assumptions about the kind of curvature allowed in our filaments. Indeed, we are only going to make a slight alteration to our ensemble of perfectly parallel filaments (Fig. 5.3), but we are going to rely on the rigor of the previous section to justify that small change.

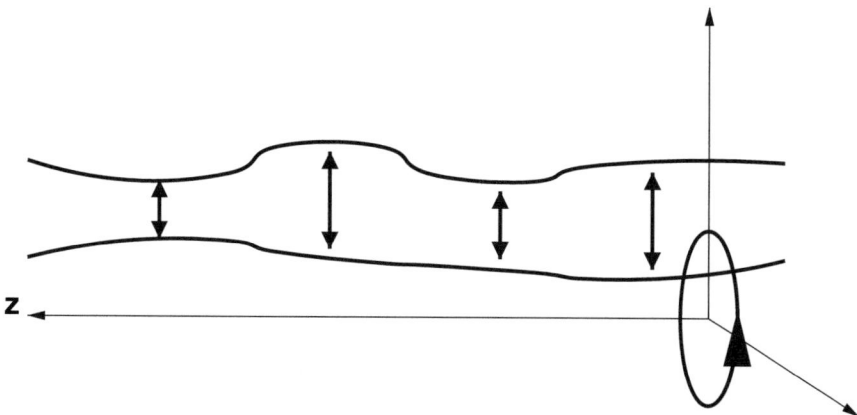

Fig. 5.3: Nearly parallel vortex filaments are filaments that have slight deviations from perfectly parallel. The Biot–Savart interaction between the filaments, therefore, is 2D and logarithmic to first order

Perfectly parallel filaments are inadequate to explain phenomena where vorticity is small and the interaction or containment forces are inadequate to keep the filaments perfectly straight. Nearly parallel vortex filaments are filaments that are quasi-two-dimensional in the sense that they vary by small amounts in the previously ignored dimension. Adding this variance in the third dimension creates finite kinetic energy at the macroscopic filament level which can change the statistics and stability of the system significantly from a perfectly parallel model.

In particular, nearly parallel vortex filaments have no negative temperature states because, with a kinetic term, their energy has no positive temperature upper bound. If you read Chap. 4, you will see that means that, by introducing a small third dimension feature, Kolmogorov's inverse energy cascade is immediately reversed! This is why, when a large-scale meteorological structure loses its small scale source of energy (e.g., a temperature gradient between warm water and cold atmosphere in the case of hurricanes) and its circulation weakens, it falls apart; all that large-scale energy comes immediately tumbling down. Because we are looking at equilibria, negative temperature states do not arise, not because they cannot on some time scale, but because by assuming a 2.5D component to the filaments we are adjusting the time scale to one where negative temperature states are no longer equilibrium states.

This re-emphasizes the point that there is no such thing as true statistical equilibrium or equilibrium statistical mechanics. Statistical equilibrium is all about the time scale you choose.

The nearly parallel formulation is the least complex quasi-2D model of vorticity and has one particularly attractive feature in that its Hamiltonian resembles that of the point vortex gas with the simple addition of a kinetic energy term. The kinetic term is a local induction approximation (LIA).

In the asymptotic analysis above, the motion of a single filament is given by

$$\frac{\partial \mathbf{X}}{\partial t} = \frac{\Gamma}{4\pi}(\log(2/\delta) + C_1)\kappa \hat{\mathbf{b}} + \mathbf{Q}_f, \qquad (5.29)$$

where $\delta = \varepsilon/R$ and $\kappa = 1/R$ is the curvature. The contribution $\mathbf{Q}_f$ is the non-local induction from the Biot–Savart integral. In the LIA, the characteristic wavelength of the filament is much smaller than the average radius of curvature, $L \ll R$, and non-local effects become small, vanishing as $R \to \infty$—the perfectly straight case. The LIA can then be written as

$$\frac{\partial \mathbf{X}}{\partial t} = C_0 \hat{\mathbf{t}} \times \frac{\partial^2}{\partial s^2}\mathbf{X}, \qquad (5.30)$$

where $\hat{\mathbf{t}}$ is the tangential vector. This uses the approximation

$$\kappa \hat{\mathbf{b}} \approx \hat{\mathbf{t}} \times \frac{\partial^2 \mathbf{X}}{\partial s^2} + O(\varepsilon), \qquad (5.31)$$

where $\varepsilon^2 = \log(1/\delta)$ is a small parameter [78].

As with the non-local induction equations, C_0 must be measured indirectly by measuring the motion of the filament and deriving its value from the equations.

If the filament is nearly parallel, then $\hat{\mathbf{t}} \approx \mathbf{e}_z$, the basis vector in the z-direction in Cartesian coordinates (x,y,z). The resulting vector then approximately points in the x-y plane and $z \approx s$ if scaled appropriately. Under this assumption, $\mathbf{X}(s,t) = (x(s,t), y(s,t))$ is two-dimensional and the LIA is

$$\frac{\partial \mathbf{X}}{\partial t} = J\left[C_0 \frac{\partial^2}{\partial s^2}\mathbf{X}\right], \qquad (5.32)$$

where

$$J = \begin{pmatrix} 0 & -1 \\ 1 & 0 \end{pmatrix} \qquad (5.33)$$

is a skew-symmetric matrix.

Although heuristically, the nearly parallel vortex filament Hamiltonian can be inferred simply by combining the Hamiltonians of a single wavy filament and the interaction term of a set of perfectly parallel filaments, a rigorous treatment shows that the system exists provided that it obeys the asymptotic constraints, $\varepsilon \ll \varepsilon^2 \ll 1$. Given a set of N nearly parallel vortex filaments with two-dimensional parameterized positions,

$$\mathbf{X}_j(s,t) = (x_j(s,t), y_j(s,t)), \tag{5.34}$$

the equations for interacting filaments with circulations Γ_k are

$$\frac{\partial \mathbf{X}_j}{\partial t} = J \left[\alpha_j \Gamma_j \frac{\partial^2}{\partial s^2} \mathbf{X}_j \right] + J \left[\sum_{k \neq j}^{N} 2\Gamma_k \frac{(\mathbf{X}_j - \mathbf{X}_k)}{|\mathbf{X}_j - \mathbf{X}_k|^2} \right] \tag{5.35}$$

The constant α_j is a core structure constant which assumes that changes in the core are slow enough relative to the given time scale, i.e., the core structure changes more slowly than the filament moves, which is a reasonable assumption since changes in the core structure correspond to decay and dissipation [29]. This constant is entirely determined, to order, by the axial mass flux (the flux of mass tangential to the filament's curve) and the initial age (the age at $t = 0$) of the core. Contrast these equations to the equations for the motion of point vortices,

$$\frac{\partial \mathbf{X}_j}{\partial t} = J \left[\sum_{k \neq j}^{N} 2\Gamma_k \frac{(\mathbf{X}_j - \mathbf{X}_k)}{|\mathbf{X}_j - \mathbf{X}_k|^2} \right], \tag{5.36}$$

which depend only on interaction.

Nearly parallel vortex filaments have a striking similarity to 2D quantum bosons with a logarithmic interaction. We will use this idea when we come back to nearly parallel vortex filaments with some of our own research in Chap. 9. Suffice it to say that nearly parallel vortex filaments' main source of interest is that they are a simple way to examine what happens when perfectly parallel vortex filaments begin to lose their inverse energy cascade, i.e., the negative temperature states that Onsager discovered [117]. In this sense, they sit between a regime of order, where 2D structures are maintained by the inverse energy cascade, and a regime of chaos where the flow becomes completely turbulent, and we will leave them there for now.

5.3 Filament Crossing and Reconnection

In most of this chapter, we have treated vortex filaments as if they were unchangeable entities. As with all entities, whether they be stars, molecules, or the Milky Way galaxy, change and the loss of distinct identity is inevitable when mergers happen.

When multiple filaments exist within a fluid, their motion can bring them into contact with one another. At this point, the asymptotic assumptions of the previous section break down and core structure becomes important, particularly viscosity.

Without viscosity, filaments cannot pass through one another at all, limiting their configurations to topologically equivalent ones. This comes from the frozen field equation,

$$\frac{\partial \omega}{\partial t} = \nabla \times (\mathbf{v} \times \omega), \tag{5.37}$$

where $\omega = \nabla \times \mathbf{v}$, which can be derived from the Euler equations for inviscid, incompressible fluid flow (and there is an equivalent theorem for magnetic flux lines). This implies that for any closed curve C the flux of the fluid is constant,

$$\oint_C \mathbf{v} \cdot d\mathbf{l} = \int_S \omega \cdot \mathbf{n} dS = \text{const.}, \tag{5.38}$$

by Alfven's theorem.

Discovering which configurations are equivalent to which is beyond the scope of this book but is related to knot theory, a branch of pure mathematics. Knot theory is, in fact, an outgrowth of attempts to answer the question of equivalent vortex filament configurations in the late nineteenth and early twentieth centuries. With viscosity, the vortex merger/reconnection process, as it is called, is a complex process best modeled by the complete Navier–Stokes equations. Kevlahan [72] has studied this process using numerical PDE solvers and a statistical description is not likely sufficient to understand it. In quantum fluids (discussed in Chap. 6), reconnection has also been studied [21].

Experimentally, vortex crossing has been demonstrated in simple "tabletop" experiments. For example, Kleckner and Irvine [76] were able to study the creation and dynamics of knotted vortices of water visually. In creating a simple trefoil knot, a knot with one crossing, they found that the vortex filament would inevitably cross itself and break, suggesting that vortex knots are highly unstable, at least in water. Presumably, a less viscous fluid would display less instability because it is more purely Euler, and it would be interesting to try the same experiment in a quantum fluid where recrossing is theoretically prohibited. Very complex vortex entanglements are also possible, especially in superfluids [17, 127].

Another important topic in non-parallel filaments is helicity, defined for a flow field **u** as

$$H = \int_V \mathbf{u}(\nabla \times \mathbf{u}) dV,$$

or the dot product of the velocity field with the vorticity integrated over the volume. Given an inviscid fluid, the helicity is conserved by the frozen nature of the vortices, i.e., lack of reconnection and merger properties. Helicity measures the degree of knottedness of a tangle [108]. Helicity is intimately related to the limiting form of the Gauss linking number (the Calugareanu invariant) [109].

5.4 Concluding Remarks

In this chapter, we have focused on the fundamentals of non-2D vortex filaments while avoiding many of the harrowing calculations associated with them. We have looked at what happens when a Navier–Stokes fluid meets an Euler fluid at the boundary of a thin filament. We have also looked at what happens when we use that knowledge to add a kinetic term to perfectly parallel filaments and how that makes the negative temperature states that first made such filaments interesting vanish. We have also briefly touched on filament merger, which is a significant subject all its own that is, mostly, beyond the scope of this book. In the following three chapters, we will examine more specific applications of vortex filaments to the subjects touched on in Chap. 2, then delve into computational methods and some of our own research.

Chapter 6
Quantum Fluids

In the previous three chapters we have focused on classical fluids. As discussed in Chap. 2, however, vortices have manifestations in quantum fluids as well, where they appear as quantized lines of vorticity. Because they are quantized, these vortex filaments are ideal for experimental studies of angular momentum, turbulence, and vorticity. Here, dissipation and viscosity are either diminished or totally absent, and, therefore, they exist in a much more idealized realm than their messy classical counterparts. Their mathematical description, however, is quite different from classical vortices because they are described by a wavefunction.

In this chapter, we discuss quantized vortex filaments in two kinds of media, Bose–Einstein Condensates and type-II superconductors (where they take the form of quantized lines of magnetism). We will examine the mathematical description of these vortices and arrive at a celebrated result by Abrikosov known as the Abrikosov state, which earned him the Nobel Prize when it was first observed some 50 years after he proposed it. It is not the purpose of this chapter to give exhaustive reviews of results in quantum fluids or even vortices in quantum fluids. For those see, [11, 16, 22, 111].

6.1 Bose–Einstein Condensates

First, we will discuss the theoretical background for Bose–Einstein Condensates.

Because bosons can occupy the same quantum states as one another, given N indistinguishable atoms, each with a set of continuous momentum states, $|k\rangle$[1] The free particle likelihood is

$$\lambda = e^{-k^2/2mT}, \qquad (6.1)$$

---

[1] The notation $|\cdot\rangle$ was invented by Paul Dirac and is called "bra-ket" notation. It denotes a continuum of quantum states indexed by a variable, e.g., k for momentum or x for position, and acts on a quantum operator such as energy similarly to how a vector acts on a matrix.

where m is the boson mass and T is the temperature. Integrating over all possible states,

$$N = V \int \frac{d^3k}{(2\pi)^3} \frac{\lambda(k)}{1-\lambda(k)}, \qquad (6.2)$$

where $1-\lambda(k)$ is the normalization of the integral over λ and V is the volume, gives a critical temperature, $T = T_c$, below which the bosons form a condensate. This temperature has the formula,

$$T_c = \left(\frac{n}{\zeta(3/2)}\right)^{2/3} \frac{2\pi\hbar^2}{mk_B}, \qquad (6.3)$$

where n is the particle density and ζ is the Riemann zeta function with $\zeta(3/2) \approx 2.6124$.

6.1.1 Modeling Quantized Vortex Lines in BECs

Quantized vortex lines can be either organized into a regular lattice (laminar) or disorganized (vortex tangles). They have a quantum of circulation,

$$\oint \mathbf{v}_s \cdot d\mathbf{l} = \frac{h}{m} = \Gamma, \qquad (6.4)$$

where h is Planck's constant and m is the helium mass. For helium $\Gamma \approx 9.97 \times 10^{-4} \text{cm}^2$. Under constant rotation with angular velocity Ω they appear as an array of ordered lines with areal (2-D) density $2\Omega/\Gamma$. These lines were first observed directly in 1972 [4] and later observed in dilute BECs (Fig. 6.1) by rotating a slightly asymmetrical magnetic trap.

The superfluid velocity around the axis of a vortex line is $v_{s,\phi} = \Gamma/2\pi r$. This diverges as $r \to 0$. Therefore, we must ask what happens there. Although the most accurate models for superfluid helium are complex couplings between the Navier–Stokes and nonlinear Schrödinger's equations, for weakly interacting atoms, such as Lithium or Rubidium, which achieve superfluidity at about 50 nK, the Gross–Pitaevskii mean-field nonlinear Schrödinger equation [62] is very accurate,

$$i\hbar \frac{\partial \psi}{\partial t} = -\frac{\hbar^2}{2m}\nabla^2 \psi + U_0|\psi|^2 \psi + V(t,r)\psi, \qquad (6.5)$$

where U_0 is the mean coupling constant, ψ is the wavefunction for the entire superfluid, and $V(t,r)$ is the potential for the rotation (an elliptical magnetic trap in the case of alkali atoms). If the wavefunction is $\psi = Ae^{i\Phi}$, velocity is given by

$$\mathbf{v}_s = \frac{\hbar}{m} \nabla \Phi. \qquad (6.6)$$

6.1 Bose–Einstein Condensates

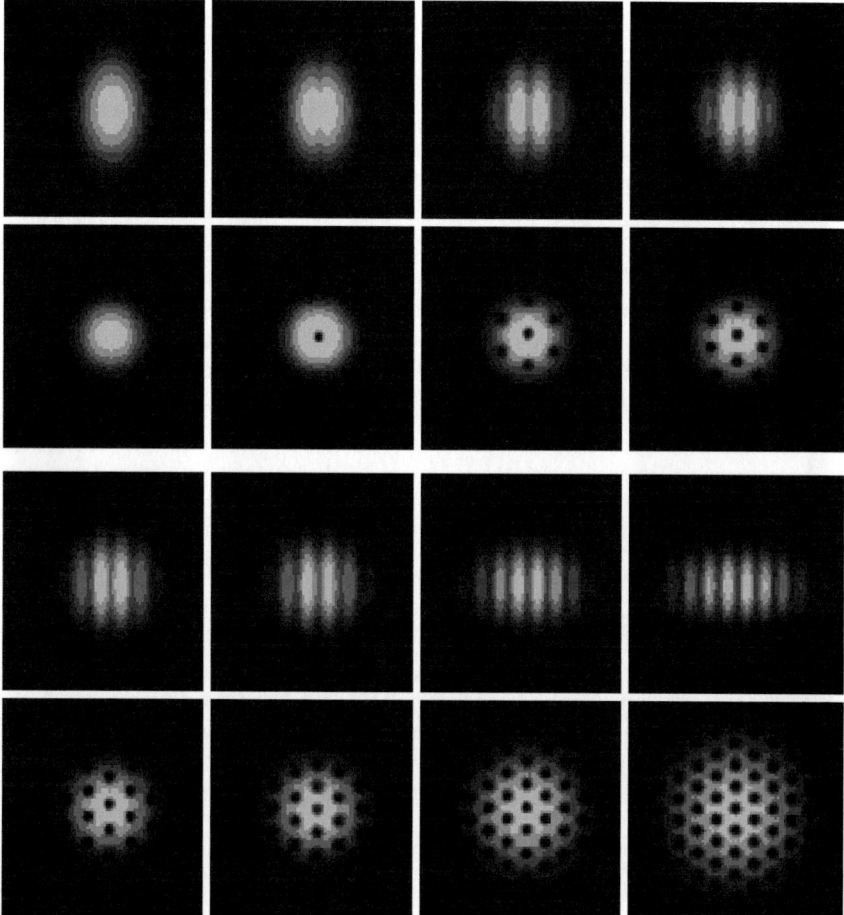

Fig. 6.1: This set of photos shows the formation of vortices in a trapped BEC from the side and top. *Courtesy of NIST*

This has a vortex solution if Φ is simply azimuth, $\Phi = \phi$. Now, from this solution ($\psi = Ae^{i\phi}$) we find that the density drops to zero from its bulk value (outside of any vortex) continuously over a distance of about $a_0 \approx 10^{-8}$cm or 1 Angstrom. Hence, there is no fluid at $r = 0$, and velocity is nominally zero there as well. Despite its shortcomings in modeling strongly interacting superfluids, this model has gained great popularity both in modeling dilute alkali Bose–Einstein condensates and Helium-II at $T = 0$ [16].

The vortex lattice structure observed in Fig. 6.1 has been beautifully simulated from the 3D Gross–Pitaevskii equations in rotation regimes that are difficult to achieve experimentally. The condensate is typically contained with a transverse angular frequency ω_t. At rotational frequencies beyond a critical value $\Omega > \Omega_c$

vortices begin to nucleate much like bubbles in a pot of water that has begun to boil. In the fast rotation regime $\Omega > \omega_t$, however, the confinement cannot hold the BEC together anymore. Experiments have achieved speeds up to $0.99\omega_t$ with the harmonic confinement alone and $1.05\omega_t$ with both harmonic and laser confinement, but we turn to simulations for rotational frequencies beyond this.

Given that we emphasize, in this text, theory that is as close as possible to experiment, we will go through a detailed simulation of [40] where real, experimental parameters determine the simulation setup. In this simulation, vortex lattices of up to 100 vortices were simulated with a quadratic plus quartic trapping potentials (approximating the harmonic plus laser confinement) with a Runge–Kutta–Crank–Nicolson numerical solver propagating the wave function in imaginary time.

Considered a BEC with $N = 3 \times 10^5$ atoms confined by a potential in cylindrical coordinates for a condensate rotating about the z-axis,

$$V(r,z) = V_h(r,z) + U(r),$$

where $V_h(r,z)$ is the magnetic potential and a laser beam propagating along the z-axis creates the potential $U(r)$. Thus,

$$V_h = \frac{1}{2}m(\omega_t^{(0)})^2 r^2 + \frac{1}{2}\omega_z^2 z^2, \quad U(r) = U_0 e^{-2r^2/w^2},$$

with trapping frequencies $\omega_t^{(0)} = 2\pi \times 75.5\,\text{Hz}$ and $\omega_z = 2\pi \times 11\,\text{Hz}$, creating a fusiform (spindle-shaped) condensate. The laser waist is $w = 25\,\mu\text{m}$ and the amplitude is $U_0 = k_B \times 90\,\text{nK}$. When r/w is small, i.e., the width of the BEC is much smaller than the waist, the potential $V(r,z)$ becomes, to first order,

$$V_1 = \left[\frac{1}{2}m(\omega_t^{(0)})^2 - \frac{2U_0}{w^2} + \frac{2U_0}{w^4}r^4 + \frac{1}{2}m\omega_z^2 z^2\right]$$

For the quadratic plus quartic potential, the transverse trapping frequency is lowered to $\omega_t = 2\pi \times 65.6\,\text{Hz}$.

The initial condition is the Thomas–Fermi distribution,

$$\rho_{TF}(r,z) = \frac{m}{4\pi\hbar^2 a_s}\left(\mu - V_1(r,z) + \frac{1}{2}m\Omega^2 r^2\right),$$

with $a_s = 5.2\,\text{nm}$ the scattering length and μ the chemical potential that satisfies the density constraint $\int d^3 r \rho_{TF} = N$. The analytical form of μ can be found for certain Ω. For high Ω the condensate becomes a prolate spheroid and nearly 100 vortices appear. For these simulations $240 \times 240 \times 240$ grid points are used which is the fine mesh needed to resolve the vortices. As Ω becomes even higher, the center of the BEC develops a funnel which becomes a hole through the center with smaller vortices appearing in the ring around the giant vortex at $\Omega/2\pi = 71$. These results go well beyond what can be achieved experimentally suggesting how simulation can probe fruitful paths for experimentalists to follow.

These simulations are a new player in vortex statistics because, until about 10–15 years ago, it was difficult to simulate up to 100 vortices even in the comparatively simpler GP equations, let alone the Navier–Stokes. With exponentially increasing speeds in high performance computing and supercomputing becoming increasingly inexpensive, however, such high resolution simulations will likely become the norm in both quantum and classical physics.

6.2 Superconductors

The Ginzburg–Landau theory represents superconductivity in terms of a complex order parameter ψ. This theory is essentially macroscopic, i.e., it does not include the microscopic mechanism of Cooper pairing that underlies superconductivity. The order parameter simply describes the phase. The theory first gives a Gibbs free energy,

$$F = F_{\text{normal}} + \alpha|\psi|^2 + \frac{\beta}{2}|\psi|^4 + \frac{1}{2m'}|(-i\hbar\nabla - 2e'\mathbf{A})\psi|^2 + \frac{\|\mathbf{B}\|^2}{2\mu_0}, \quad (6.7)$$

where F_{normal} is the free energy of the normal, non-superconducting phase, α, β are phenomenological parameters, and e', m' are effective charge and mass parameters. In this case $e' = 2e$, the charge of the Cooper pair and $m' = 2m_e$. This implies that in the normal phase (unmixed) $\psi = 0$. Minimizing the free energy with respect to the order parameter and the vector potential, $\mathbf{A}$, gives the stationary Ginzburg–Landau equations:

$$\frac{1}{2m'}(-i\hbar\nabla - \frac{e'}{c}\mathbf{A})^2\psi + \alpha\psi + \beta|\psi|^2\psi = 0,$$

$$\nabla \times (\nabla \times \mathbf{A}) = -\frac{4\pi}{c}\mathbf{j}_s, \quad (6.8)$$

$$-\frac{ie'\hbar}{2m'}(\psi^*\nabla\psi - \psi\nabla\psi^*) - \frac{e'^2}{m'c}|\psi|^2\mathbf{A} = \mathbf{j}_s,$$

where $\mathbf{j}_s$ is the superconducting electric current density and the total current is $\mathbf{j} = \mathbf{j}_s + \mathbf{j}_n$ [57]. This has the boundary condition at the superconductor-vacuum interface,

$$\mathbf{n}\left(-i\hbar\nabla - \frac{e'}{c}\mathbf{A}\right)\psi = 0, \quad (6.9)$$

where $\mathbf{n}$ is the normal to the interface surface. The superconducting behavior is defined by the parameters,

$$\lambda_0 = \sqrt{\frac{m'c^2\beta_c}{4\pi e'^2|\alpha|}}, \quad \varkappa = \frac{m'c}{e'\hbar}\sqrt{\frac{\beta_c}{2\pi}} = \frac{\sqrt{2}e}{\hbar c}H_{\text{cm}}\lambda_0^2, \quad (6.10)$$

where $\beta_c \equiv \beta(T_c)$. Here λ_0 is the penetration depth for weak magnetic fields $H \ll H_{cm}$ and H_{cm} is the critical magnetic field for massive samples [56].

Quantized flux lines in superconductors are also vortices, called London vortices. Given a magnetic field $\mathbf{B} = (0, 0, B_z)$, we have a differential equation,

$$\frac{d^2 B_z}{dr^2} + \frac{1}{r}\frac{dB_z}{dr} - \frac{B_z}{\lambda^2} = 0, \tag{6.11}$$

where

$$\lambda = \frac{m_e}{\mu_0 n_s e^2}, \tag{6.12}$$

is the London penetration depth, m_e is the electron mass, n_s is the local density of superconducting carriers, and e is the electron charge.

This is a form of Bessel's equation with solutions that are modified or hyperbolic Bessel's functions, $K_\nu(z)$. The solution to this one is proportional to $K_0(z)$ and the resulting field is [10]

$$B_z(r) = \frac{\Phi_0}{2\pi\lambda^2} K_0(r/\lambda), \tag{6.13}$$

where Φ_0 is the magnetic flux enclosed by the vortex core,

$$\Phi_0 = \oint \mathbf{A} \cdot d\mathbf{l} = \frac{h}{2e}. \tag{6.14}$$

The quantized flux is very small, $\Phi_0 \approx 2 \times 10^{-15}$ Wb. For small values where $\lambda \gg r$, $K_0(r/\lambda) \sim -\ln(r/\lambda)$ and

$$B_z(r) = -\frac{\Phi_0}{2\pi\lambda^2} \ln(r/\lambda). \tag{6.15}$$

From the magnetic field, the self-energy of a single flux line is

$$E_1 = -\frac{\Phi_0^2}{2\pi\lambda^2} \ln\left(\frac{\xi}{\lambda}\right), \tag{6.16}$$

where ξ is the coherence length (size of a Cooper pair) of the superconductor. See Fig. 6.2 for a diagram of a cross-section of a London vortex. A similar formula appears in the self-energy of vortex filaments in the Navier–Stokes approximation [77, 144]. It does not, however, include interaction with other filaments which is responsible for the following:

To obtain the interaction force and energy, we begin with the free energy:

$$\mathscr{F} = \frac{1}{2\mu_0} \int \left[B^2 + \lambda^2 |\nabla \times \mathbf{B}|^2 \right] dV$$

where $B^2 = |\mathbf{B}|^2$ is the magnetization and $\lambda^2 |\nabla \times \mathbf{B}|^2$ is the kinetic energy (from the GL equations) and the equivalence for two vortices at r_1 and r_2:

6.2 Superconductors

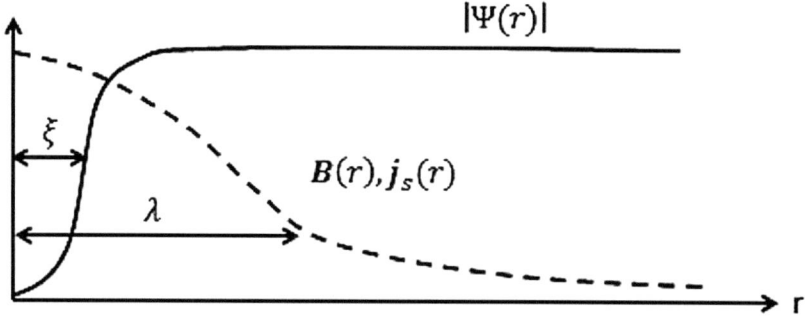

Fig. 6.2: Radial dependence of the order amplitude and magnetic field as a function of radial distance from the vortex center

$$B^2 + \lambda^2 |\nabla \times \nabla \times \mathbf{B}| = \Phi_0[\delta(r-r_1) + \delta(r-r_2)].$$

The free energy, thus, becomes

$$F = \frac{\Phi_0}{2\mu_0}(B(r_1) + B(r_2)).$$

Let

$$B(r_1) = B(0) + B_{12}(x), \quad B(r_2) = B(0) + B_{12}(x).$$

Then

$$F = 2E_1 + \frac{\Phi_0}{2\mu_0} 2B_{12}(x).$$

Thus, the energy of interaction is

$$U(x) = \frac{\Phi_0}{\mu_0} B_{12} = \frac{\Phi_0^2}{2\pi\lambda^2} K_0\left(\frac{x}{\lambda}\right),$$

with a force of

$$f = -\frac{dU}{dx} = -\frac{\Phi_0}{\mu_0} \frac{B_{12}}{dx} = -\Phi_0 j_{12}(x),$$

where $j_{12}(x)$ is the current. In the limit of $\lambda \gg x$ we have the familiar vortex–vortex interaction of Euler fluids:

$$U(x) \approx \frac{\Phi_0^2}{2\pi\lambda^2} \ln\left(\frac{x}{\lambda}\right).$$

The Abrikosov lattice exists in a phase between two critical external field strengths $H_{c1} < H < H_{c2}$. For $H < H_{c1}$ there is insufficient field strength for a single vortex to form. For $H > H_{c2}$ the lattice's intervortex spacing approaches the vortex core size ξ and a second order phase transition to the normal state takes place. The first critical field is found from (6.16) to be

$$H_{c1} = \frac{E_1}{\Phi_0}$$

while the second critical field is given by calculating the field when the intervortex distance is $\sim \xi$ which turns out to be:

$$H_{c2} = \frac{\Phi_0}{2\pi\mu_0\xi^2}.$$

The vortex glass is a disordering transition in the GL equations, and one of the most fascinating phenomena discovered in the last couple decades. Numerical simulations have shown initially strong turbulent states evolve into spirals which grow until the entire space is filled. Each spiral has a four or five sided polygonal structure [11]. This glass phase occurs in the $H > H_{c1}$ phase (above the Meissner or vortex free phase) but below the vortex liquid phase that happens above a certain temperature $T_g(H)$ that depends on H [49]. Numerical simulations have confirmed this phase diagram [148]. In high-T_c superconductors fluctuation effects are very important because of the melting of the vortex lattice (solid). The glassy phase is an example of a symmetry breaking of the long-range order. This leads to a high nonlinear resistivity, $\rho(j) \sim e^{-j_t/j)^\mu}$ with $\mu \in [0,1]$ and j_t the threshold current. The linear resistivity, however, vanishes meaning the system is truly superconducting [111]. Weak disorder in the lattice results in a Bragg glass which retains the order of the lattice. Bragg glasses have been observed experimentally via neutron diffraction showing a power-law decay of the vortex crystal order [79]. This phase yields to the vortex glass which is strongly topologically disordered.

Since we want to focus on seminal results and simple but insightful calculations, we are going to focus on the mapping onto the 2D XY model. XY models are often used to model systems with an order parameter such as superfluids and superconductors because their phase transitions are not always accompanied by a symmetry breaking. This was first proposed by Fisher [49]. Fisher's approach was to take a nearly parallel model (different from the one for fluids in Chap. 5) with vortices parameterized by z, $r_i(z) = (x_i(z), y_i(z))$ and map the vortex model onto a dual continuum boson model:

$$H_0 = \int d^2r \left[-\frac{T}{2\varepsilon_1} \Phi^\dagger(r)\nabla_r^2\Phi(r) + \frac{1}{4\pi}(H_{c1}-H)\hat{B}(r) \right], \tag{6.17}$$

$$H_1 = \frac{1}{16\pi^2\lambda^2} \int d^2r d^2r' \hat{B}(r) K_0\left(\frac{r-r'}{\lambda}\right)\hat{B}(r'), \tag{6.18}$$

$$H_p = \int d^2r V_p(r,z)\hat{B}(r), \tag{6.19}$$

which are the London self-energy, the vortex–vortex interaction, and the pinning energy, respectively. Also, $\hat{B}(r) = \Phi_0\Phi^\dagger(r)\Phi(r)$, $\Phi_0 = hc/2e$, $\varepsilon_1 = H_{c1}\Phi_0/4\pi$, and K_0

6.2 Superconductors

is the modified Bessel's function. The duality between bosons and vortex filaments allows Fisher to propose a model for pinned vorticity with a random (Gaussian) pinning potential $V_p(r,z)$ with mean zero and variance Δ appropriate for microscopic pinning defects (oxygen defects) in bulk superconductors.

Fisher's model develops an effective replicated Hamiltonian via ensemble averaging of the nearly parallel model. In other words, Fisher employs the so-called replica trick. This is a trick for pulling averages into logarithms by replicating our system n times. For example, let the system have a partition function Z. Then n replicas has a partition function Z^n. Since $Z^n = e^{n \ln Z} = 1 + n \ln Z + O(n^2)$ we know that $\langle \ln Z \rangle = \lim_{n \to 0} \frac{1}{n} (\langle Z^n \rangle - 1)$ or sometimes for convenience $\ln \langle Z^n \rangle = n \langle \ln Z \rangle + O(n^2)$. This trick allows Fisher to average over the Gaussian pinning potential and develop an interacting replica model,

$$H_n = \sum_{\alpha=1}^{n} H_{0,\alpha} + H_{1,\alpha} - \frac{\Delta}{T} \int d^2 r \sum_{\alpha,\beta} \hat{B}_\alpha(r) \hat{B}_\beta(r),$$

where α and β sum over the replicas. He shows then that the replica model has a glassy phase which he then extends to $n \to 0$ by tossing out a dimension and using a 2D XY model.

A slightly different XY model treatment due to Nattermann illustrates the derivation [111]. Throwing out a dimension leaves us with a superconducting film, basically a slice of a bulk conductor where the external field is parallel to the film. In this case, the phase diagram has $H_{c1} = \frac{2}{\pi} \frac{\Phi_0}{s^2} \ln s/\xi$ and $H_{c2} = \frac{\sqrt{3}}{\pi} \frac{\Phi_0}{\xi s}$ where s is the thickness of the film. We assume that $\lambda > s/\pi > \xi$. These films are experimentally feasible and vortices can be observed in them. Next, assume that the superconducting film is in the (x,z) plane with the vortices parameterized by the z axis (nearly parallel along one dimension). We have an average line spacing a. Let the nth vortex be given by $x_n(z) = X_n + u_n(z)$ where $X_n = na$ and $u_n(z) = u(X_n, z)$ are the mean position and the displacement from mean, respectively. The Hamiltonian is $H = H_0 + H_p$,

$$H_0 = \int d^2 r \frac{c_{11}}{2} (\partial_x u)^2 + \frac{c_{44}}{2} (\partial_z u)^2, \tag{6.20}$$

$$H_p = \int d^2 r \rho_u(r) V_p(r), \tag{6.21}$$

where $\rho_u(r)$ is the line density, c_{11} is the compression, and c_{44} is the tilt elastic constant. Let $\langle V_p(r) \rangle = 0$ and $\langle V_p(r) V_p(r') \rangle = \Delta(x - x') \delta(z - z')$ where $\Delta(x) \approx \Delta_0/(\sqrt{2\pi}\xi) e^{-x^2/2\xi^2}$ and $\Delta_0 = n_i^{(2)} f_p^2 \xi^3 s$ where $n_i^{(2)}$ is the impurity density per area and f_p is the force of an impurity on a vortex line. Given vortex lines with a stiffness $\varepsilon_{\text{stiff}}$ and a short-ranged, nearest neighbor interaction energy $U(a + u_{n+1} - u_n)$, then $c_{11} = aU''(a)$ and $c_{44} = \varepsilon_{\text{stiff}}/a$.

Thermal fluctuations less than the intervortex spacing lead to a repulsion between the lines. From the XY model we can also derive the glass transition temperature $T_g = (c_{11} c_{44})^{1/2} a^2/\pi$. (For details see [111] and the references therein.)

What we have shown above is a model for the transition from an ordered vortex lattice to a disordered vortex glass. Like an ordinary glass, the glass phase transition is not like melting or freezing but a semi-dynamical disordered state. The critical take-away from this chapter is to note how the vortex lattice can be obtained in quantum fluids such that, under certain constraints, mathematically, it is identical to the lattice in classical fluids. The difference, however, is that in quantum fluids vortices are strictly quantized whereas they can take on any circulation in classical fluids. In the next chapter, we leave the realm of the very cold for the very hot with plasmas where vortex filaments make yet another appearance and show, once again, how fundamental they are to nature.

Chapter 7
Plasmas

7.1 Magnetohydrodynamics

For decades nuclear fusion has promised to be a source of nearly limitless energy, but, because of technical problems, it lags far behind other sources of energy in the race to replace fossil fuels. In the Sun fusion occurs because of the spherically symmetric pressure of the Sun's self-gravity. This type of fusion is relatively stable because each particle experiences this gravitational attraction individually and is individually pulled toward the center. Therefore, significant numbers of particles, compared to the Sun's overall bulk, cannot escape the Sun's pull and stop the fusion process. (A fraction do continuously achieve escape velocity creating the Solar wind.)

Although recent successes with fusion reactors such as the Joint European Torus (JET) have increased the promise of fusion power as a viable source of energy in this century, recreating Solar conditions in the laboratory is fraught with difficulty because instabilities in magnetically confined plasmas disrupt fusion and cause repeated energy loss. When deuterium fuses with tritium, they produce a helium nucleus (alpha particle), a high energy neutron, and energy,

$$^2_1D + ^3_1T \rightarrow ^4_2He + ^1_0n + 17.6 \text{MeV}. \tag{7.1}$$

Because the protons within each nucleus tend to repel one another strongly, however, activation energy is much higher than non-fusion reactions (although lower than other fusion reactions), and optimal reaction temperature is approximately 100 million Kelvin. At these temperatures, plasma particles have enormous pressure and require an equally large counter pressure to sustain a reaction.

Effective confinement is essential to attaining a stable reaction and overcoming the proton repulsion and plasma pressure. Gravity is too weak at small scales to be useful for containment. One of the most advanced alternatives involves a tokamak where strong electric ring currents and a toroidal forcing current magnetically confine the superheated plasma within toroidal containment area. The combination of a poloidal magnetic field, **B**$_P$, from the forcing current through the center of the torus and a toroidal magnetic field, **B**$_T$, from currents run around the outside at intervals along the torus, forces the plasmas into helical lines or filaments that are also nearly parallel where the ions or electrons orbit around the force lines. Given a sufficiently large major axis, the radius of curvature is large enough that a two-dimensional Biot–Savart interaction governs their thermodynamic behavior, and the lines are nearly parallel. These lines can then be modeled as if they were particles in their own right with their own interaction properties. The guiding center model of electron plasmas is an example [70].

In theory if the magnetic field is strong enough it can compress the electrically conducting plasma sufficiently to overcome the plasma's self-repulsion and force a fusion reaction. Unexplained instabilities arise that disrupt the fusion reaction however [151]. At the time of writing, fusion can only be sustained for a few seconds. Although models such as that of Kadomtsev have accounted for plasma instabilities in small reactors, the instability problem is far from solved in large reactors such as JET nor has even a satisfactory qualitative explanation been reached. Plans for a larger and more stable fusion reactor (ITER) and commercial reactors in the decades ahead demand a deeper understanding of the underlying causes of instability in magnetically confined plasmas.

Plasmas are magnetohydrodynamic, two fluid systems. The coupled equations for a magnetohydrodynamic fluid are

$$\frac{d\rho}{dt} + \rho \nabla \cdot \mathbf{v} = 0, \tag{7.2}$$

$$\rho \frac{d\mathbf{v}}{dt} + \nabla p - \mathbf{j} \times \mathbf{B} = \mathbf{0}, \tag{7.3}$$

$$\mathbf{E} + \mathbf{v} \times \mathbf{B} = \mathbf{0}, \tag{7.4}$$

$$\frac{d}{dt}\left(\frac{p}{\rho^\Gamma}\right) = 0, \tag{7.5}$$

where ρ is the density, p is the pressure, $\mathbf{j}$ is current, $\mathbf{v}$ velocity, and $\Gamma = 5/3$ is the ratio of specific heats [50]. These are also the equations of motion for an inviscid, adiabatic, perfectly conducting, and electrically neutral liquid, but in the context of plasmas these equations are only valid under the conditions,

$$v_{\text{therm}}/\delta \gg \bar{v} \gg \delta v_{\text{therm}}, \tag{7.6}$$

where $v_{\text{therm}} = \sqrt{2k_B T/m}$ is the average thermal velocity, $\bar{v} = \langle \|\mathbf{v}\| \rangle$ is the average fluid velocity, and δ mean ratio of the particle gyro-radius (also called the Larmor radius) to the length scale of motion. As we will see, the ratio δ is similar to the ratio we saw in superconductors, ξ/λ, in that it relates vortex core sizes to typical length scales.

7.2 Confined 2D Plasmas

The equations of motion for a magnetic field perpendicular to the (x, y) plane are

$$q\frac{dx_i}{dt} = \frac{1}{B}\frac{\partial H}{\partial y_i}, \quad q\frac{dy_i}{dt} = -\frac{1}{B}\frac{\partial H}{\partial x_i}, \tag{7.7}$$

for particles of like charge q, where the Hamiltonian is

$$H = \sum_{i<j} -\frac{2q^2}{l} \ln|\mathbf{r}_i - \mathbf{r}_j|, \tag{7.8}$$

l is the length of the filaments, and $\mathbf{r}_i = (x_i, y_i)$. (Note that the filaments could have different charges on them, q_i and q_j, but this makes the problem much more difficult and is not necessary to the current demonstration.) This two-dimensional charged vortex fluid is the same as the equations for interacting parallel vortex lines in an incompressible inviscid fluid.

Without going into the full derivation which can be found in [46], one can show that in a microcanonical ensemble of N vortices in a fixed volume container of volume V that the system satisfies the following equation of state,

$$\frac{1}{T} = \frac{1}{Nq^2}\left[\exp\left(\frac{-E}{Nq^2}\right) - \frac{3}{2}\right], \tag{7.9}$$

where $n = N/V$ is the density, which implies that there is a critical energy $E_m = -\ln(3/2)Nq^2$ where, for $E > E_m$, temperature is negative.

7.3 Quasi-2D Electron Columns

While the 2D guiding center model has negative temperature states, the quasi-2D model does not. This is because the guiding center model is a short-time scale equilibrium state while the quasi-2D model applies to longer time-scales. At these time scales, negative temperature states are not stable equilibria. A new phenomenon, however, appears in quasi-2D plasmas with long-range interactions that is not present in the 2D model: negative specific heat. While [134] has proven that systems that are not isolated from the environment must have positive specific heat, the specific heat in isolated systems can be negative [94]. Negative specific heat is an unusual phenomenon first discovered in 1968 in microcanonical (isolated system) statistical equilibrium models of gravo-thermal collapse in globular clusters [95]. In gravo-thermal collapse, a disordered system of stars in isolation undergoes a process of core collapse with the following steps: (1) faster stars are lost to an outer halo where they slow down, (2) the loss of potential (gravitational) energy causes the core of stars to collapse inward some small amount, (3) the resulting collapse causes the stars in the core to speed up. If one considered the "temperature" of the cluster to be the average speed of the stars, this process has negative specific heat because a loss of energy results in an increase in overall temperature.

In a magnetic fusion system or other thermally isolated plasma, should negative specific heat exist, the related runaway collapse could have profound implications for fusion where extreme confinement is critical to a sustained reaction.

The guiding center plasma or ideal fluid vorticity model for quasi-2D columns of electrons or lines of vorticity is similar to the widely studied two-dimensional model but the lines contain small variations which can change the dynamics of the system. An ensemble of filaments $\{\psi_1(\tau), \ldots, \psi_N(\tau)\}$,

$$E_N[\psi_1 \ldots \psi_N] = \sum_i \int_0^l d\tau \frac{\alpha \Gamma}{2} \left| \frac{d\psi_i}{d\tau} \right|^2 - \frac{1}{\varepsilon} \sum_{i<j} \Gamma^2 \log|\psi_i - \psi_j|, \qquad (7.10)$$

where $1/\varepsilon$ is the coupling constant, all the filaments have the same average circulation, Γ, α is a core elasticity constant related to the frequency, and l is the length of the period under periodic boundary conditions $\psi_i(0) = \psi_i(l)$. Each filament $\psi_i(\tau) = (x_i(\tau), y_i(\tau))$ is a vector in the plane with a parameter τ representing the third dimension. This is under special asymptotic assumptions that the filaments are nearly parallel and far enough apart [92]. An example of such an ensemble is pictured in Fig. 9.1.

There is a significant difference between the 2D one-component Coulomb plasma and this quasi-2D model. Because the total energy is entirely dependent on how far apart the lines are, an ensemble of two-dimensional lines at a fixed energy has a maximum radius beyond which the lines cannot move while the quasi-2D lines can move, theoretically, as far apart as there is space available because the potential energy can always be balanced by altering the kinetic energy.

Statistical derivations for three-dimensional fluids of any kind have tended to focus on single vorticity columns [29, 65] and [77, 144] or statistical treatments of ensembles of two-dimensional point vortices [34, 46, 70, 117]. The nearly parallel vortex filament model of [78, 92] for Navier–Stokes fluids is an exception.

The nearly parallel model can be extended to electron columns by a generalized vorticity model for electron plasmas [61, 74, 147] which takes the magnetic and electric fields into account as well as the vorticity. In the case of charged particles, the vorticity must, essentially, be gauge invariant: the magnetic field, $-e\mathbf{B}/m = (-e/m)\nabla \times \mathbf{A}$, and the charged fluid vorticity, $\omega = \nabla \times \mathbf{v}$, combine into a general vorticity field $\Omega = m^{-1}\nabla \times \mathbf{p}$ where the generalized or "canonical" gauge invariant momentum is $\mathbf{p} = m\mathbf{v} - e\mathbf{A}$, m is the electron mass, $-e$ is the electron charge, $\mathbf{v}$ is the fluid velocity field, and $\mathbf{A}$ is the magnetic vector potential field. The generalized angular momentum is $\mathbf{L} = \mathbf{r} \times \mathbf{p}$.

Electron column core sizes are small, equal to the Larmor radius, δ, about a tenth of a millimeter for electrons, hence the core structure may be abstracted by the local induction approximation (LIA). The size of the local region, which is the wavelength of the highest energy frequencies on the filaments, is naturally given by the London wavelength, $b = c/\omega_{pe}$, where the electron plasma frequency is given

7.3 Quasi-2D Electron Columns

by $\omega_{pe}^2 = ne^2/m\varepsilon_0$ and n is the electron number density of the individual filament [147]. Because the filaments are all nearly parallel, collisions between points not in the same cross-section are small, and the two-dimensional Coulomb interaction gives the potential to leading order.

A filament nearly parallel to the z axis has a C^2 curve $\psi(\tau) = (x(\tau), y(\tau))$ with $L^2_{[0,l]}$ derivative where $l \ll 1$ is a small length scale. Assuming no short-wave disturbances (knots, kinks, etc.), the filament has an LIA kinetic energy functional (derived from 5.8),

$$T_1[\psi] = \int_0^l d\tau \frac{\alpha \Gamma}{2} \left| \frac{d\psi}{d\tau} \right|^2 \tag{7.11}$$

where Γ is the generalized circulation of the filament [147] and $\alpha = \log(b/\delta) + 1$, b is the arclength of the "local" region where the induction takes place, i.e., the wavelength of the highest frequency modes on the filament, and δ is the core size of the filament. The ratio b/δ does not vary significantly over the length of the filament; therefore, it can be assumed to be constant [144]. The derivation of LIA can be found in [126].

For N such filaments, they have, under suitable asymptotic assumptions such that the filament core size is much smaller than the intervortex spacing, a 2-D Coulomb interaction such that

$$V_N[\psi_1 \ldots \psi_N] = \sum_i \int_0^{l_i} d\tau - \frac{1}{\varepsilon} \sum_{i<j} \Gamma^2 \log |\psi_i - \psi_j|, \tag{7.12}$$

where all the filaments have the same average circulation [74]. The kinetic energy is the square amplitude of the filament with zero amplitude implying zero kinetic energy; hence, the filament's self-energy is not directly connected to the microscopic temperature. Combining the kinetic and interaction energies gives us $E_N = T_N + V_N$ or (7.10).

The kinetic energy of generalized angular momentum for a single filament of electrons is $A_1 = \frac{1}{2} I \bar{\omega}^2$ where I is the moment of inertia of the filament, $I = \int_0^l d\tau |\psi|^2$, and $\bar{\omega}$ is the generalized frequency of rotation. For N filaments, all with the same frequency,

$$A_N[\psi_1 \ldots \psi_N] = \frac{1}{2} \mu' \sum_i \int_0^{l_i} d\tau \Gamma |\psi_i|^2, \tag{7.13}$$

where $\mu' \Gamma = \bar{\omega}^2$. Now choose mass and charge units such that $e = m = 1$ for electrons.

No boundary conditions are imposed perpendicular to z since the plasma is fully contained by the magnetic field and never contacts any surfaces [35, 78, 90, 92, 112]. The confinement radius R is a bounding radius representing the distance from the center of the plasma to the outer edge where the density falls to zero (sometimes abruptly). This is smaller than the radius of the container. Provided the magnetic surfaces to which they are confined are closed, the filaments cannot interact with material surfaces.

7.3.1 A Mean-Field Approach

Our goal is to find the specific heat of this vortex model in statistical equilibrium given an appropriate definition for energy and a microcanonical (isolated) probability distribution. Our approach is to describe the statistical behavior of a large number of discrete, interacting vortex structures and consider the limiting case.

We start with a finite number of filaments and take the limit, $N \to \infty$, later, keeping total vorticity, $\Lambda = \int_{\Re^3} \Omega(\mathbf{r}) d\mathbf{r}$, constant by rescaling. This is known as a **non-extensive thermodynamic limit** because the overall vortex strength stays constant even as the number of vortex filaments increases towards infinity.

The generalized angular momentum is

$$M_N = \sum_{i=1}^{N} \Gamma \int_0^1 d\tau |\psi_i(\tau)|^2. \tag{7.14}$$

From this point on, we assume the vortex circulations are all scaled to unity, $\Gamma = 1$.

For our isolated, classical system, the energy and angular momentum plus magnetic moment are conserved giving rise to the following probability distribution for the filaments in equilibrium:

$$P(s) = Z^{-1} \delta(NH_0 - E_N - pM_N) \delta(NR^2 - M_N), \tag{7.15}$$

where H_0 is the total enthalpy per vortex per period of the plasma, s is the complete state of the system, and $Z = \int ds \, \delta(NH_0 - E_N - pM_N) \delta(NR^2 - M_N)$ is a normalizing factor called the partition function. Here E_N is the energy functional and M_N is the angular momentum. It is our intent to allow R^2, the size of the system,

$$R^2 = \lim_{N \to \infty} \left\langle N^{-1} \int_0^1 d\tau |\psi_i(\tau)|^2 \right\rangle, \tag{7.16}$$

to be determined by other parameters in the system and keep enthalpy and pressure, p, fixed.

The size of the configuration space (partition function), Z, cannot be found in closed-form by any known analytical methods. Since our aim is an explicit expression for specific heat, we need to find a closed-form approximation of Z.

To simplify the equations, we combine the large number of vortices into two average or "mean" vortices, which results in a mean vorticity field. This is the "mean-field" approach common in statistical mechanics. Our mean vortices are as follows: one mean vortex is a mean distance from the origin. The other is the statistical center of charge of all the filaments—a single, perfectly straight filament fixed at the origin with strength of the remaining vortices, $N - 1 \sim N$.

Given a filament i and a filament j, the mean-field approximation implies the following:

$$\langle |\psi_i - \psi_j| \rangle \longrightarrow \sqrt{\int_0^1 d\tau |\psi_i(\tau)|^2} = ||\psi_i||, \tag{7.17}$$

7.3 Quasi-2D Electron Columns

where i is any filament index and the double bars, $||\cdot||$, indicate $\mathscr{L}_2$-norm on the interval $[0,1]$.

The energy function now reads as follows:

$$E'_N = \int_0^1 d\tau \sum_{i=1}^N \left[\frac{\alpha}{2} \left|\frac{\partial \psi_i}{\partial \tau}\right|^2 - \frac{N}{4}\log||\psi_i||^2 \right]. \tag{7.18}$$

This assumption makes all vortices are statistically independent, and the statistics of all the vortex structures can be found from those of one. We modify (7.15) and (7.14) appropriately and drop primes.

In statistical mechanics of isolated systems, all equilibrium statistics can be determined from maximizing the entropy. The entropy per filament, S_N, is defined by

$$e^{S_N} = \int D\psi \delta(NH_0 - E_N - pNR^2)\delta(NR^2 - M_N). \tag{7.19}$$

This definition implies

$$S_N = \log\left[\int D\psi \delta(NH_0 - E_N - pNR^2)\delta(NR^2 - M_N)\right]. \tag{7.20}$$

We have now set the stage to describe our derivation of negative specific heat.

Given the space available, we proceed to outline, rather than fully derive, our method of obtaining an explicit formula for the maximal entropy of this mean-field system in the non-extensive thermodynamic limit (as defined above) from which we obtain an explicit, closed-form formula for the specific heat.

First it is important to note that our approach relies heavily on the steepest-descent methods in [69] and spherical model approach in [20, 64, 71, 141]. The works of [86, 90] have preceded and inspired this work in their novel applications of the spherical model to barotropic vorticity models on the sphere. These works lay the ground for our derivation.

The steepest-descent and consequently spherical model methods convert Dirac-delta functions into their Fourier-space equivalent integral representations. The procedure for our derivation relies on two closely related facts: the integral representation of the Dirac-delta function in a microcanonical distribution, for example:

$$\delta(Nx - Nx_0) = \int_{\beta_0 - i\infty}^{\beta_0 + i\infty} \frac{d\beta}{2\pi i} e^{N\beta(x-x_0)}, \tag{7.21}$$

and the steepest-descent limit, again only an example:

$$\lim_{N\to\infty} \frac{1}{N} \int_{\beta_0 - i\infty}^{\beta_0 + i\infty} d\beta e^{N\beta(x-x_0)} = e^{\beta_0(x-x_0)}, \tag{7.22}$$

which allows quantities such as entropy to be determined, provided we can determine what β_0 is, a method for determining which [69] provides.

Using the first fact, we can simplify the microcanonical problem by converting delta functions into integrals and re-ordering the phase-space (ψ) and parameter-space (β) integrals to arrive at

$$e^{S_N} = \int \frac{d\beta}{2\pi i} e^{\beta N H_0} Z_{\text{bath}}, \qquad (7.23)$$

where

$$Z_{\text{bath}} = \int D\psi \, e^{-\beta E_N - \mu M_N} \delta(NR^2 - M_N) \qquad (7.24)$$

is a semi-microcanonical partition function with a microcanonical angular momentum constraint. In the infinite N limit, Z_{bath} approaches the canonical partition function (i.e., the same function but without the Dirac-delta factor in the integrand of Eq. 7.24), making their use interchangeable in the limit.

The maximal entropy per filament per period, having the form,

$$S_{\max}(H_0) = \lim_{N \to \infty} N^{-1} S_N, \qquad (7.25)$$

is where the limit comes into play.

Coulomb interactions have a problem in that as the number of "particles" (in this case vortex filaments) grows infinitely large, the interaction energy grows with the square of the number of filaments. The solution is to rescale the temperature, which in turn, rescales the interaction energy. Rescaling the temperature causes a chain of necessary scalings to restore the balance so that other quantities do not go to zero: $\beta' = \beta N$, $\alpha' = \alpha/N$, $p' = p/N$, and $H_0' = H_0/N$. These are reasonable because only the interaction energy needs rescaling.

With all these scalings there are no more mathematical obstructions, and the maximal entropy can be found by the standard mathematical procedures in [20, 69]. We provide only the final formula obtained in view of space constraints, but the procedure is quite straightforward once the appropriate framework is set up:

$$S_{\max}(H_0) = \beta_0' H_0' + \frac{\beta_0'}{4} \log(R^2) - \frac{1}{2\alpha \beta_0 R^2} - \beta_0' p' R^2. \qquad (7.26)$$

where

$$R^2 = \frac{\beta_0'^2 \alpha' + \sqrt{\beta_0'^4 \alpha'^2 + 32 \alpha' \beta_0'^2 p'}}{8 \alpha' \beta_0'^2 p'}, \qquad (7.27)$$

where the mean temperature, β_0, is as yet unknown. This entropy is exact within the mean-field assumption for $N \to \infty$.

By [69], we find the unknown multiplier, β_0, by relating the enthalpy per filament parameter, H_0, to the mean enthalpy, $NH_0 = \langle E_N + pM_N \rangle$, where $\langle \cdot \rangle$ denotes average against Eq. (7.15).

7.3 Quasi-2D Electron Columns

By Eq. (7.15) the average enthalpy is given by

$$\langle E_N + pM_N \rangle = \frac{\int D\psi (E_N + pM_N)\delta(NH_0 - E_N - pM_N)\delta(NR^2 - M_N)}{\int D\psi \delta(NH_0 - E_N - pM_N)\delta(NR^2 - M_N)}, \quad (7.28)$$

and, again going through some steepest-descent-based calculations given in [69], we find the formula,

$$H_0' = \frac{\partial}{\partial \beta_0'} \left(-\frac{\beta_0'}{4} \log R^2 + \frac{1}{2\alpha'\beta_0'R^2} + \beta_0' p' R^2 \right) \quad (7.29)$$

exactly. We cannot give an explicit expression for β_0 because it is a root of a transcendental equation, but such is unnecessary for the following negative specific heat result:

We define specific heat at constant generalized pressure, p,

$$c_p = -\beta_0^2 \frac{\partial H_0}{\partial \beta_0} \quad (7.30)$$

and after evaluating with Eq. (7.29) and simplifying (dropping primes and 0-subscripts)

$$c_p = \frac{\beta}{4} \left(\frac{\alpha \beta^2}{\sqrt{\alpha \beta^2 (\alpha \beta^2 + 32p)}} - 1 \right). \quad (7.31)$$

Equation (7.31) is significant. It indicates that the specific heat is not only negative for this system, but *strictly* negative if parameters are non-zero (Fig. 7.1). In the low-temperature (large β) case, for constant field strength, R^2 does not change significantly with temperature indicating that filaments are in a stable configuration for a large range of low-temperatures. Because the filaments do not move relative to one another at low-temperatures and the self-induction is negligible, the enthalpy does not change. As temperature rises, the increase in internal entropy causes a massive expansion in the overall size of the system. The strong magnetic field absorbs this energy, but, since it is assumed to be an infinitely massive reservoir able to maintain the enthalpy at H_0, the confinement remains constant.

Only a local-induction approximation is necessary for this analysis, even though 2D point vortices have no negative specific heat in equilibrium. The negative specific heat here can be explained as a process: (1) a vortex's Brownian motion causes it or part of it to move away from the center, (2) potential energy decreases, (3) the vortices in the center can move closer together and temperature increases.

The negative specific heat indicates a runaway reaction (i.e., the fixed energy, fixed angular momentum equilibrium point is metastable). Considering its similarity to gravo-thermal collapse: we hypothesize that the metastable point could have a collapse similar to globular clusters in which an outer halo of columns separates from an inner core that collapses in on itself, possibly resulting in nuclear fusion.

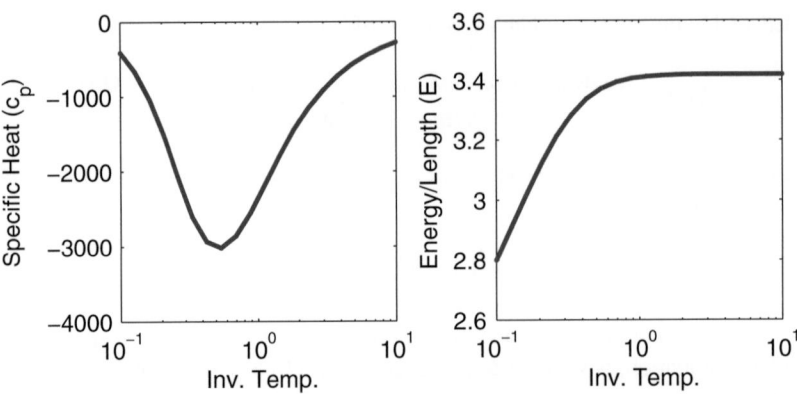

Fig. 7.1: The specific heat at constant pressure (Eq. 7.31) for the thermally isolated system is negative, meaning that the constant pressure Enthalpy Per Length (Eq. 7.29) decreases with increasing temperature. (Here $\alpha' = 5 \times 10^5$ and $p' = 8 \times 10^4$.)

7.3.2 A Variational Approach

We now turn to the variational method as an alternative approach to showing negative specific heat. In the thermodynamic limit as $N \to \infty$ such that total circulation, $\Lambda = \Gamma N$, is constant, the filaments have area density $f(\mathbf{c}, \mathbf{r}, \tau) = \Gamma g(\mathbf{c}, \mathbf{r}, \tau)$ (where $\mathbf{c} = d\psi/d\tau$ and Γ is the generalized circulation) such that, if $\sigma = (\mathbf{c}, \mathbf{r})$, the kinetic energy is

$$\mathscr{T} = \frac{1}{2} \int d^4\sigma \, d\tau \, f \alpha c^2, \qquad (7.32)$$

where $c = \|\mathbf{c}\|$. In a rotating frame the kinetic energy of angular momentum becomes a potential added to the interaction potential,

$$\mathscr{V} = \frac{1}{2}\mu' \int d^4\sigma \, d\tau \, f |\mathbf{r}|^2 - \frac{1}{\varepsilon} \int d^4\sigma \, d^4\sigma' \, d\tau \, ff' \log|\mathbf{r} - \mathbf{r}'|, \qquad (7.33)$$

where $f' = f(\mathbf{c}', \mathbf{r}', \tau)$ and particle number is $N = \int d^4\sigma \, d\tau \, g$. All integrals are over the interior of the torus. Since the energy functional does not depend on τ, the density with maximal entropy does not depend on τ either (as one can show from the variation); therefore, we drop the integrals over τ and assume the area density is constant in τ. All functionals are now per unit length.

For a fixed energy system, the entropy (with Boltzmann's constant $k_B = 1$),

$$S = -\int d^4\sigma \, g \log g, \qquad (7.34)$$

is maximal in the most-probable macrostate.

7.3 Quasi-2D Electron Columns

To maximize the entropy, we must solve the variational problem,

$$\delta S = 0, \tag{7.35}$$

subject to the constraints, $\mathcal{T} + \mathcal{V} = E$, N fixed. That is for some small parameter λ, we define a family of density functions $g(\sigma;\lambda)$, such that entropy is maximal ($\delta S = dS/d\lambda = 0$) at $g(\sigma;0)$.

The method of Lagrange multipliers provides the equation for the variation of S subject to the constraints,

$$\delta S + \beta' \delta E + \nu \delta N = 0, \tag{7.36}$$

where β' is inverse temperature, ν is a normalization parameter. (Angular momentum is automatically conserved by the rotational invariance of the energy.) Taking the variation [95] we have

$$\log g + 1 + \beta' \Gamma \left(\frac{\alpha}{2} c^2 + \frac{\mu'}{2} r^2 + \psi \right) + \nu = 0, \tag{7.37}$$

where

$$\psi(\mathbf{r}) = -\frac{1}{\varepsilon} \int d\mathbf{r}' \rho(\mathbf{r}') \log |\mathbf{r} - \mathbf{r}'| \tag{7.38}$$

is the 2D Coulomb potential. Let $H = \frac{\alpha}{2} c^2 + \frac{\mu'}{2} r^2 + \psi$.

Solving (7.37) for the density,

$$g = A e^{-\beta' \Gamma H}, \tag{7.39}$$

where $A = \exp[-(\nu + 1)]$ is a normalization constant that gives the particle number N. Let $\beta = \beta' \Gamma$. Showing the equipartition of energy, the average kinetic energy per filament is

$$\frac{\int d^2 c f \frac{\alpha}{2} c^2}{\int d^3 c f} = \frac{\int d^2 c \exp\left(-\frac{1}{2} \beta \alpha c^2\right) \frac{1}{2} \alpha c^2}{\int d^2 c \exp\left(-\frac{1}{2} \beta \alpha c^2\right)} = \frac{1}{\beta}. \tag{7.40}$$

To solve for the spatial density, ρ, we can integrate Eq. (7.39) over all "velocities", $\mathbf{c}$,

$$\rho = \int d^2 c f = B \exp[-\beta(\psi + \mu' r^2/2)], \tag{7.41}$$

where $B = A(2\pi/\beta)$.

Replacing ρ in the potential with the above equation gives an integral equation for the most-probable potential inside the circle,

$$\psi(\mathbf{r}) = -\int d\mathbf{r}' B e^{-\beta(\psi(\mathbf{r}') + \mu' r'^2/2)} \log |\mathbf{r} - \mathbf{r}'|. \tag{7.42}$$

This integral equation is equivalent to the Poisson equation,

$$\nabla^2 \psi(\mathbf{r}) = -\frac{4\pi}{\varepsilon}\rho = \begin{cases} -4\pi B \exp[-\beta(\psi + \frac{\mu'}{2}r^2)]/\varepsilon & |\mathbf{r}| < R \\ 0 & |\mathbf{r}| \geq R \end{cases} \quad (7.43)$$

with boundary conditions such that ψ and $d\psi/dr$ are continuous at the boundary $r = R$. Because of the axisymmetry of the energy the potential must also be statistically axisymmetric. Converting to polar coordinates and (7.43) gives the ODE,

$$\frac{1}{r}\frac{d}{dr}\left(r\frac{d\psi}{dr}\right) = -\frac{4\pi}{\varepsilon} B e^{-\beta(\psi(r) + \mu'r^2/2)}, \quad (7.44)$$

for $r < R$.

Because the potential is repulsive, the only solutions to this equation have finite density everywhere (unlike gravitational systems, infinite densities such as black holes are not possible in a repulsive Coulomb system). To simplify, we make the change of variables, $v_1 = \beta(\psi - \psi(0))$ and $r_1 = \sqrt{4\pi\beta A \exp[-\beta\psi(0)]/\varepsilon} \cdot r = \sqrt{4\pi\beta\rho(0)/\varepsilon} \cdot r$, which simplifies the ODE (7.44) to

$$\frac{d^2 v_1}{dr_1^2} + \frac{1}{r_1}\frac{dv_1}{dr_1} + e^{-v_1 - \beta\mu' r_1^2 (R^2/z^2)/2} = 0, \quad (7.45)$$

where $z = \sqrt{4\pi\rho(0)\beta/\varepsilon}R$ and with boundary conditions,

$$v_1(0) = v_1'(0) = 0, \quad (7.46)$$

where $v_1' = dv_1/dr_1$. Note that the plasma density can be written $\rho = \rho(0)e^{-v_1}$ or ($v_1 = -\log[\rho/\rho(0)]$); therefore, the variable v_1 describes how the density changes as the distance from the origin changes and decreases monotonically from 0 at $r_1 = 0$.

The viral theorem of Clausius relating kinetic and potential energy to pressure [38, 95] may be applied,

$$2\mathscr{T} + \mathscr{V} = 3pV, \quad (7.47)$$

where

$$p = \int_{r=R} d^2 c f \frac{1}{3}\alpha c^2, \quad (7.48)$$

is the surface "pressure" and $V = \pi R^2$ is the area of the circle. From the equipartition theorem, where Λ is the total circulation,

$$\mathscr{T} = \frac{\Lambda}{\beta}. \quad (7.49)$$

Using the Virial theorem,

$$E = 3pV - \mathscr{T} = 3pV - \frac{\Lambda}{\beta}, \quad (7.50)$$

7.3 Quasi-2D Electron Columns

and

$$\mathcal{V} = 3pV - 2\frac{\Lambda}{\beta}. \tag{7.51}$$

Evaluating the surface pressure,

$$p = \int_{r=R} d^2c\, f \frac{1}{3}\alpha c^2 = \frac{2}{3}\frac{1}{\beta}\int_{r=R} d^2c\, f = \frac{2\rho(R)}{3\beta}, \tag{7.52}$$

where

$$f = A\exp\left[-\beta\left(\alpha\frac{c^2}{2} + \mu'\frac{r^2}{2} + \psi\right)\right]. \tag{7.53}$$

Integrating (7.44), circulation Λ is given by

$$\frac{\Lambda}{\varepsilon} = -\left(r\frac{d\psi}{dr}\right)_{r=R} = -\frac{1}{\beta}\left(r_1\frac{dv_1}{dr_1}\right)_{r_1=z}. \tag{7.54}$$

Then

$$\beta = -\frac{\varepsilon z v_1'(z)}{\Lambda}. \tag{7.55}$$

For the rest of this section, v_1' and v_1 shall refer to $v_1'(z)$ and $v_1(z)$ only.

The pressure is

$$p = \frac{2}{3}\frac{\rho(R)}{\beta} = \frac{2}{3}\frac{\rho(0)e^{-v_1 - \beta\mu R^2/2}}{\beta}. \tag{7.56}$$

Since $r_1^2/r^2 = |4\pi\beta\rho(0)/\varepsilon|$, and eliminating instances of β with (7.55),

$$p = \frac{2}{3}\frac{\varepsilon z^2}{4\pi R^2}\frac{e^{-v_1 - \beta\mu' R^2/2}}{\beta^2} = \frac{\Lambda^2}{6\pi R^2\varepsilon}\frac{z^2 e^{-v_1 + \mu' z v_1' \varepsilon R^2/(2\Lambda)}}{(-zv_1')^2}. \tag{7.57}$$

Let $\mu = \varepsilon\mu' R^2/(2\Lambda)$, then

$$3pV = \frac{\Lambda^2}{2\varepsilon}\frac{z^2 e^{-v_1 + \mu z v_1'}}{(-zv_1')^2}, \tag{7.58}$$

and the energy from (7.50) is

$$E = \frac{\Lambda^2}{\varepsilon}\left(\frac{z^2 e^{-v_1 + \mu z v_1'}}{2(-zv_1')^2} - \frac{1}{(-zv_1')}\right). \tag{7.59}$$

The entropy, obtained from (7.34), is given by

$$S = \frac{1}{\Gamma}\left\{\beta\left(E - \frac{\Lambda^2}{\varepsilon}\log R\right) - \Lambda\log p\beta^2 + \Lambda\log 2\pi\Gamma\right\}. \tag{7.60}$$

The specific heat is

$$c_v = \frac{dE}{dT} = \frac{\frac{dE}{dz}}{\frac{dT}{dz}}. \quad (7.61)$$

When $\mu z v_1' r^2 / R^2 = v_1(z)$ we have a constant density solution, where the potential energy and the confinement are perfectly balanced. For $\mu z v_1' r^2 / R^2 > v_1(z)$ the potential dominates, and the density tends to favor expansion with higher density toward the wall of the container. These profiles, although stable, are not suitable for containment. For $\mu z v_1' r^2 / R^2 < v_1(z)$ the magnetic confinement dominates and a more Gaussian density profile is preferred with higher density toward the middle. These profiles are useful for containment.

Numerically evaluating v_1 and v_1' for a range of z values (Fig. 7.2) in the strong rotation regime $\mu > 0.5$, we have only negative specific heat. In the weak rotation regime with $\mu < 0.5$, the specific heat has positive states which are stable (Fig. 7.3).

Entropy maximization of the model yields the following results:

1. An expression for the energy in terms of z, $v_1(z)$, and $v_1'(z)$,

$$E = \frac{\Lambda^2}{\varepsilon} \left(\frac{z^2 e^{-v_1 + \mu z v_1'}}{2(-zv_1')^2} - \frac{1}{(-zv_1')} \right), \quad (7.62)$$

where Λ is the total circulation, ε is the coupling constant for vorticity, and μ is a parameter determining strength of the angular kinetic energy;

2. An expression of the inverse temperature of the filaments (Boltzmann's constant, $k_B = 1$),

$$\frac{1}{T} = \beta = -\frac{\varepsilon z v_1'(z)}{\Lambda}; \quad (7.63)$$

3. An expression for the central density,

$$\rho(0) = \frac{\varepsilon z^2}{4\pi \beta R^2}, \quad (7.64)$$

where R is the radius of the confinement area;

4. A second ODE governing the density at a particular energy/temperature,

$$\frac{d^2 v_1}{dr_1^2} + \frac{1}{r_1} \frac{dv_1}{dr_1} + e^{-v_1(r_1) + \mu r_1^2 v_1'(z)/z} = 0; \quad (7.65)$$

with boundary conditions,

$$v_1(0) = v_1'(0) = 0, \quad (7.66)$$

where $v_1' = dv_1/dr_1$;

5. And an expression for the density:

$$\rho(r) = \rho(0) \exp(-v_1(r) + \mu r^2 z v_1'(z)/R^2), \quad (7.67)$$

where $r = r_1 / \sqrt{4\pi \beta \rho(0)/\varepsilon}$ is the distance from the center.

7.3 Quasi-2D Electron Columns

From numerical evaluation of (7.65) the energy (7.62), inverse temperature (7.63), and core density (7.64) are computed as functions of z. These are plotted for several values of μ (Figs. 7.2, 7.3 and 7.4).

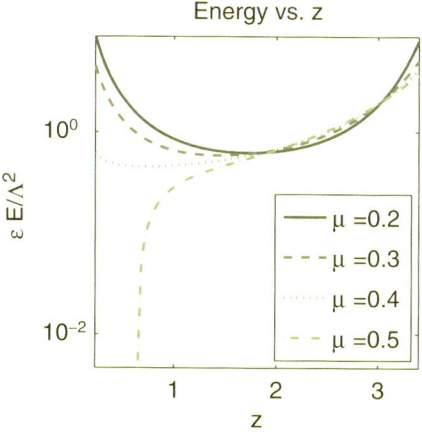

Fig. 7.2: For regimes with $\mu < 0.5$, confinement is weak enough that stable states exist, indicated by energy decreasing with z, but beyond that all states are metastable, indicating that equilibrium is not possible. When energy increases with z, the system is metastable, and, because this leads the system to run away to hotter temperatures where the central density decreases, runaway expansion results

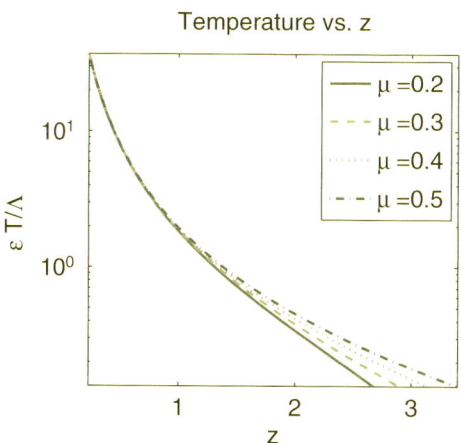

Fig. 7.3: Temperature decreases with z. Therefore, if energy increases with z, the specific heat is negative. If energy decreases with z, the system has stable equilibria

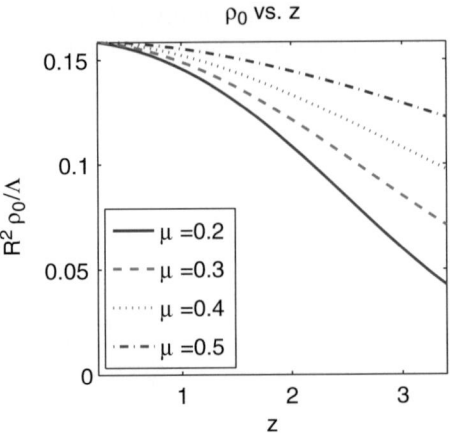

Fig. 7.4: As μ increases, confinement becomes stronger and the central density decreases less with increasing z, but, because it is decreasing in the metastable regime, this indicates a core expansion

7.4 Interpretation of Negative Specific Heat

Negative specific heat indicates a metastable state where the entropy, rather than being globally maximal, is at a saddle point or local maximum, and the system evolves out of this state by increasing temperature, where the term "temperature" refers to the amount of kinetic energy per unit circulation and not the electron temperature. Because the specific heat is negative, increasing temperature lowers the total energy. In gravitational systems where this phenomenon was first described, this results in gravo-thermal catastrophe where a globular cluster, for example, experiences a core collapse. The collapse of the core results in reduced gravitational potential and increased stellar velocities. Thus, the overall energy decreases while the temperature increases. It has been argued by analogy that negative specific heat in magnetically confined, neutral electron-positron plasmas would result in core collapse leading to the possibility of nuclear fusion as columns of electrons and ions collapse into one another [73]. In a paper, however, we showed that self-energy causes an anomalous expansion in the mean radius of the density profile and developed a mean-field formula for it, confirming it with Monte Carlo simulations [8]. The same approach in a microcanonical ensemble shows that the system exhibits negative specific heat [6]. Therefore, while magnetically confined quasi-2D plasmas exhibit negative specific heat indicating metastable energy states where no equilibrium exists, because the potential is repulsive, the instability is a runaway expansion, not core collapse.

This expansion arises because increases in kinetic energy and decreases in potential energy both cause expansion. Analyzing Figs. 7.2, 7.3, and 7.4, the mechanism works as follows: (1) The potential energy decreases. If the magnetic confinement is sufficiently weak at the given density, expanding it decreases the potential energy. (2) The kinetic energy increases in response, but, because the system has expanded,

7.4 Interpretation of Negative Specific Heat

energy is lost at the outer edges of the plasma. Thus, the filament temperature has increased but energy decreased. (3) The increased kinetic energy causes the density profile to spread (by increasing the variance), causing further expansion, and returning the cycle to step (1). As the potential energy continues to decrease, the kinetic energy continues to increase, and the system experiences a runaway expansion directly analogous to the runaway collapse of a gravitational system.

Unlike in the gravitational systems, however, because the specific heat is negative for all energy values, no equilibrium states exist and the plasma never settles. Because the system is forced, like a forced snow-pile (with continuous snow fall), as long as the metastable state persists, the energy input into the system from the externally applied currents balances the energy loss. When the metastable state ceases, however, the rate of energy loss increases dramatically and the rate of energy input no longer balances the rate of energy loss. This expansion causes a dramatic loss of core electron temperature which has been observed in experiment [149], a direct result of heat transport from the core to the outer edge via the runaway expansion mechanism. Recovery occurs when the filaments expand far enough apart that the rate and magnitude of their collisions decreases, and the expansion slows. Supposing that complete plasma disruption does not occur, the input current allows the energy to increase again, returning the plasma to a metastable state. Hence, the instability is regular and repeated.

This instability is a direct result of the confinement as Fig. 7.2, where confinement is weak, shows. In the strong confinement regime, expansion decreases the potential because $v_1(r)$ increases from the core showing that density (7.67) decreases; when confinement is weak, however, expansion increases it because $v_1(r)$ decreases from the core, and any further expansion increases the density at the edge of the confinement area. In a real system, weakly confined plasmas do not persist for long before complete disruption however, so this regime is not useful for sustained fusion.

Negative specific heat has been found in neutral, two-component, electron-positron plasmas [73]. This model, however, ignores Coulomb interactions between charged particles and does not take into account that the positively charged ion lines move far more slowly than the negatively charged electron lines. For a neutral electron/ion plasma, a two-fluid magnetohydrodynamical model is appropriate. At 100 million degrees K (the minimum target of tokamak reactors although we are not assuming a toroidal geometry here) electrons travel at a mean velocity of 40,000 kps while deuterons travel at "only" 600 kps. Instabilities in tokamak plasmas such as the sawtooth instability may be related to the metastable state described in this chapter [151].

A great deal more needs to be done to investigate negative specific heat is long-range interacting vortices as well as runaway reaction. One of the most promising ways to make magnetically confined nuclear fusion a reality is to understand how metastable states can be avoided as columns are brought closer together. Ideally, we want stable states with positive specific heat, and it may be that some additional confining potential or some anisotropy in the potential could provide that. These remain open problems.

Chapter 8
Computational Methods

Computational methods have been applied to statistical mechanics and combinatorics since the beginnings of electronic computers (e.g., ENIAC) in the 1940s. In this chapter, we give a brief overview of the different computational methods that have been applied to ensembles of vortex filaments, including Numerical PDEs, Monte Carlo and the Metropolis algorithm, Feynman–Kac Path Integral methods, the Demon algorithm, and Hamiltonian flow. Each of these methods has different advantages and disadvantages that we will explore.

8.1 Numerical PDEs

Numerical Partial Differential Equation (PDE) solver methods are perhaps the most widely used numerical methods in physics. In terms of accuracy, they are unparalleled because they model as closely as possible the microscopic behavior of the system under study. Thus, they have become the gold standard in simulation and modeling.

The most basic and straightforward approach to studying vortex filaments with numerical PDEs is by modeling the relevant equations, e.g., the Navier–Stokes, directly. Only recently, however, has it become possible to simulate large numbers of vortex filaments this way. High resolution simulations of (magneto)hydrodynamic turbulence, for example, requiring billions of grid points are now being performed on supercomputers (Fig. 8.1).

One of the problems with simulating vortex filaments with numerical PDEs from first principles equations such as the Navier–Stokes, Gross–Pitaevskii, Ginzburg–Landau, or the coupled Magnetohydrodynamic equations (depending on the application) is that meshes need to be fine enough to resolve slender filament structures and their core behavior. At the same time, intervortex regions need comparatively coarser meshes. One solution is to adapt the size of the mesh so that it is only fine where needed. This technique has proved successful in simulating

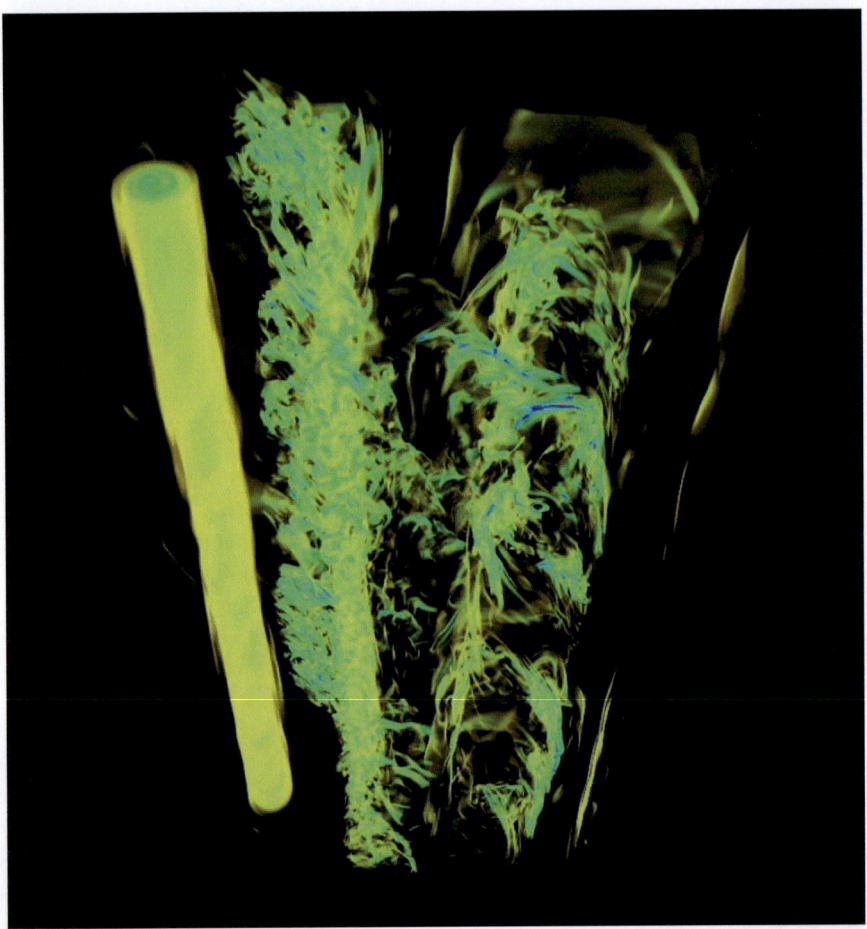

Fig. 8.1: This is a simulation of vortex filaments under rotating shows laminar flow on the *left* and tangles of small vortices on the *right*. [37] *Courtesy of CISL/NCAR/UCAR/NSF under the Creative Commons License:* `http://creativecommons.org/licenses/by-nc/3.0/legalcode`

stationary states (equilibria) of fast-rotating BECs, generating dense Abrikosov lattices [41]. This approach saves considerable computational time necessary for modeling highly interacting systems and strong rotation/external field relevant to certain experiments.

Although numerical PDEs are essential for modeling some flows, they lack the scalability of Monte Carlo methods, and, in many cases, they are overkill when a simple statistical model will provide more insight.

Because entire texts are typically devoted to numerical PDEs, we will not describe the methods in detail and, instead, move on to other methods that are relevant to the deep ocean model presented in the next chapter.

8.2 Canonical Ensemble

8.2.1 Monte Carlo

Monte Carlo algorithms are essential to statistical physics and can be traced to Ulam and Von Neumann in the 1940s for use on combinatorial problems [45]. The essence of Monte Carlo is to approximate a statistical distribution by sampling a related distribution, e.g., uniform or normal, where an exact sampling formula exists, and then by applying an acceptance test to the resulting samples. The advantage over numerical PDEs comes as the number of dimensions, i.e., degrees of freedom, increases. Numerical PDE solvers tend to be quick and effective at lower dimensions but as the degrees of freedom becomes hundreds, thousands, or more the computational complexity increases cubically. Monte Carlo methods cut through this explosion by sampling the degrees of freedom, keeping the computational complexity linear. In other words, for an increase from $N = 1,000$ degrees of freedom to $N = 10,000$, a numerical PDE increases by 1,000 times in complexity while Monte Carlo only increases by 10 times.

Monte Carlo integration, for example, allows the partition function of a system to be approximated if its energy is bounded from below (which is not always the case with vortex systems). For example, suppose we have a system with N particles, with d-dimensional states $\mathbf{x}_1, \ldots, \mathbf{x}_N$. The system has a likelihood distribution $\lambda(\mathbf{x}_1, \ldots, \mathbf{x}_N)$ and normalization function:

$$\mathcal{N}_N = \int d\mathbf{x}_1 \cdots d\mathbf{x}_N \lambda(\mathbf{x}_1, \ldots, \mathbf{x}_N),$$

such that $p = \lambda/\mathcal{N}_N$ is the probability distribution.

For $N \gg 10$ numerical integration techniques are extremely intensive. Suppose we want to calculate the partition function for $N = 1,000$, however, or $N = 10^6$? Monte Carlo integration allows us to sample the integrand rather than divide it up in some trapezoidal way. We can begin our Monte Carlo sampling by sampling a point from the uniform measure:

$$\int_D d\mathbf{x}_1 \cdots d\mathbf{x}_N,$$

with some simple bounding box or hypersphere D. We also sample a height, y, above the dN-dimensional domain from an interval $[a,b]$. (How you know what a and b are is dependent on the problem. Typically, we have some idea of maximum and minimum values for the likelihood function or can calculate them via extremal methods.)

Now, for each point,$(\mathbf{x}_1, \ldots, \mathbf{x}_N, y)$ we sample, we evaluate λ. If $\lambda \geq y$, we accept it, if $\lambda < y$, we reject the point.

After sampling many millions or billions of points (depending on the dimensions), we can count the number of accepted points, A, and take the ratio of this number to the total number of sampled points, M and this will approximate the ratio of the partition function to the hypervolume of either $D \times [a,b]$ for the uniform sampling. Since the latter are known, the normalization function can be found.

This all seems very convenient until we put it into practice on the Boltzmann distribution, $e^{-\beta E} = \lambda$. Most states have energies much larger than the most likely state. Hence, the Boltzmann distribution assigns very small probabilities to those states. This means we are going to end up sampling many states that contribute very little to the final result and potentially ignoring many states that contribute the most. In general, we are also not interested in knowing the numerical value of the normalization function anyway. There is, in fact, a better way, an algorithm that can "learn" to approximate statistical system directly so that we can calculate things that we actually care about, like energy and size. This algorithm is sometimes called Markov Chain Monte Carlo but is also known as the Metropolis algorithm after its inventors.

An interesting exercise is to implement a computer program to calculate the area of a circle with Monte Carlo. Scale the computer program up to higher dimension n-spheres. Develop a plot of the areas and compare it to the formula $V_n(R) = \frac{\pi^{n/2}}{\Gamma(n/2+1)} R^n$ where Γ is the gamma function. How well does it do? It should do fairly well. Now, try the same program to evaluate the simple Boltzmann distribution $\lambda = \exp -\beta |\mathbf{x}|^2$ for increasing dimensions of the n-vector $\mathbf{x}$. Compare it to the formula $(\beta \pi)^{n/2}$. What happens as you increase dimensions? It will do much worse than the first program.

8.2.2 Metropolis

Nicolas Metropolis, Arinna Rosenbluth, Marshall Rosenbluth, Augusta Teller, and Edward Teller first developed in 1953 what would be come the first Markov Chain Monte Carlo (MCMC) algorithm, known as the Metropolis algorithm [105]. According to the history, this began as a conversation over dinner. They were looking for a way to sample Boltzmann distributions and obtain averages such as

$$\langle F \rangle = Z^{-1}[T] \int d\mathbf{x}_1 \cdots d\mathbf{x}_N F(\mathbf{x}_1, \ldots \mathbf{x}_N) \exp -E(\mathbf{x}_1, \ldots, \mathbf{x}_N)/kT,$$

where

$$Z[T] = \int d\mathbf{x}_1 \cdots d\mathbf{x}_N \exp -E(\mathbf{x}_1, \ldots, \mathbf{x}_N)/kT,$$

for a set of N particles in $\mathbb{R}^2$. For this distribution, standard Monte Carlo fails because the Boltzmann distribution is too small for most configurations. In other words, Monte Carlo is not "smart" enough to stick to the most likely states and gets stuck in highly improbable ones. The dimensions of the problem are too large and the likely states too specific for Monte Carlo to solve in a reasonable time. This problem is called "the curse of dimensionality." This occurs in numerous other fields such as combinatorics, machine learning, and data mining. In all these areas,

8.2 Canonical Ensemble

the phenomenon is the same: the volume of a search space increase exponentially with dimension while the available samples increases at a slower pace, e.g., linearly.

The Metropolis algorithm avoids the curse by sampling in the most statistically significant parts of the search space only. The intuition behind the algorithm is to propose a random walk where the particles move, one at a time, to randomly selected positions. After each move, the new configuration accepted or rejected based on the energy of the new configuration. For example, for a single step, given a particle at (x_i, y_i), let the new *proposed* position be

$$x'_i = x_i + \sigma \xi_{1i}, \quad y'_i = y_i + \sigma \xi_{2i},$$

where ξ_{1i} and ξ_{2i} are uniform in $\mathscr{U}(-1,1)$ and σ is some step size. Now the difference in energy between the new configuration and the current one, ΔE, is computed and the proposed position is accepted with probability $\min\{1, \exp -\Delta E/kT\}$. Otherwise, the original position is retained. In other words, if the energy decreases, then the configuration is automatically accepted. If the energy increases, it is accepted with probability equal to the ratio of the Boltzmann distributions of the two states. The authors of the original algorithm also proved the validity of the algorithm by proving ergodicity, i.e., broadly speaking, that the system has the same probability distribution for one particle over N units of time as N particles over space. Hastings later generalized the Metropolis algorithm into a broad statistical tool [66], and the algorithm is frequently called the Metropolis–Hastings.

The power of this method over Monte Carlo is that once the Markov Chain has reached an equilibrium state where it has "forgotten" its initial conditions, the random walk will tend to remain in the most-probable macrostate rather than getting stuck in improbable macrostates, defeating the curse of dimensionality.

8.2.3 Path Integral Methods

The Metropolis algorithm has been successful in computing features of 2D point vortex ensembles since the 1970s [33]. In the early days of statistical computing into the 1980s and 1990s, however, the Metropolis algorithm could generally only handle classical point particle systems, but as the results in these areas became more and more esoteric or problems became harder and harder to solve and, concurrently, computing power increased exponentially, research turned from classical to quantum physics. In particular, work on new states of matter such as Bose–Einstein Condensates and Type-II superconductors led to the application of Monte Carlo methods to solving high dimensional Schrödinger equations modeling individual atoms, e.g., bosons, to confirm and expand on the results based on mean-field methods from earlier decades.

Path Integral Monte Carlo methods emerged from the path integral formulation invented by Dirac that Richard Feynman later expanded [152], in which particles are conceived to follow all paths through space. One of Feynman's great contributions to the quantum many-body problem was the mapping of path integrals onto a classical system of interacting "polymers" [48]. D. M. Ceperley used Feynman's convenient piecewise linear formulation to develop his PIMC method which he successfully applied to He-4, generating the well-known lambda-transition for the first time in a microscopic particle simulation [30]. Because it describes a system of interacting polymers, PIMC applies to classical systems that have a "polymer"-type description like nearly parallel vortex filaments.

PIMC has several advantages. It is a *continuum* Monte Carlo algorithm, relying on no spatial lattice. Only time (length in the z-direction in the case of vortex filaments) is discretized, and the algorithm makes no assumptions about types of phase transitions or trial wave-functions.

PIMC is appropriate for ensembles that have an energy functional,

$$\mathcal{H}_N = \sum_{i=1}^{N} \alpha \left| \frac{\partial \psi_i}{\partial \tau} \right|^2 + V(\psi_1, \ldots, \psi_N),$$

where $\psi_i(\tau) = x_i(\tau) + i y_i(\tau)$ is a complex number representing a filament or quantum path of a boson through time such that $\psi(0) = \psi(1)$, α is an elasticity multiplier related to the boson's mass, and V is some interaction between the bosons. This is appropriate for ^{4}He.

The core of the PIMC algorithm is bisection—the sampling of filaments by cutting them into finer and finer linear segments (Fig. 8.2).

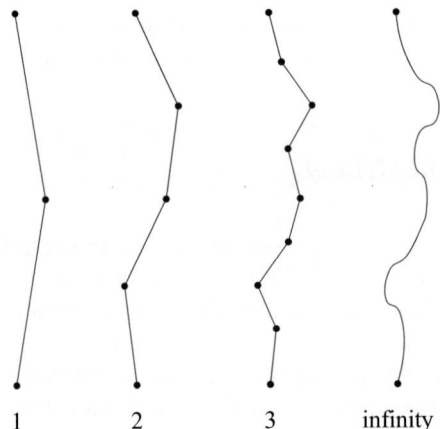

Fig. 8.2: The bisection algorithm works by bisecting the filament to sample point positions. First the center point is selected, then the two points half-way from the center to the end-points, then four more points, eight, and so on until some maximum number of points are sampled. These are snapshots of one filament at different steps in the sampling process. Each time the filament is subdivided the new center points are selected from a normal distribution

Let each filament, i, be divided into a piecewise linear vector ψ_i with M distinct points, $\psi_i(1),\ldots,\psi_i(M)$. Let $\psi_i(M+1) = \psi_i(1)$. The Monte Carlo simulation begins with a random distribution of filament end-points in a square of side, e.g., 10, and there are two possible moves that the algorithm chooses at random:

1. Moves a filament's end-points, $\psi_i(1)$ and $\psi_i(M+1)$. The index i is chosen at random, and the filament i's end-points moved a uniform random distance.
2. Keeps end-points stationary and, following the bisection method of Ceperley (Fig. 8.2), grows a new internal configuration for a randomly chosen filament [30].

The bisection method builds the filament by selecting points. It starts by selecting the mid-point between the two fixed end-points $\psi_i(1)$ and $\psi_i(M+1)$. Assume $M = 2^m$ where m is the desired bisection level. First, we select the position of $\psi_i(M/2)$ by sampling it from a normal distribution with mean $\mu = (\psi_i(1) + \psi_i(M+1))/2$ and variance $\sigma^2 = 1/\alpha$. For the second bisection step, we sample $\psi_i(M/4)$ and $\psi(3M/4)$ with means $(\psi_i(1) + \psi_i(M/2))/2$ and $(\psi_i(M/2)/2 + \psi(M+1))/2$ and variance $\sigma^2 = 1/2\alpha$. In general, the variance will decrease by half for each bisection and the mean will always be the linear interpolation between the neighboring points on the filament.

Since the bisection algorithm samples the kinetic part of the energy, the acceptance/rejection step of the Metropolis algorithm only calculates the potential energy. Let V_N^s be the potential energy of the configuration $\Psi = \psi_1,\ldots,\psi_N$ called state s. After each move, the energy of the new state, s', is calculated and retained with probability

$$A(s \to s') = \min\left\{1, \exp\left(-\beta[V_N^{s'}(M) - V_N^s(M)]\right)\right\}. \tag{8.1}$$

The stopping criteria are determined when the means of the desired observables, e.g., average energy, average angular momentum, etc. stop changing. Typically, this involves testing the average values periodically, e.g., averaging over moves 100,000–110,000 and comparing them to averages over moves 110,000–120,000. If those averages have changed less than a small threshold, then the simulation stops. Typically, convergence takes longer when the simulation parameters put the system close to a critical point, however, and the threshold must be chosen so that it does not stop prematurely. In the next chapter, we will go into a more detailed example from our own research.

8.3 Microcanonical Ensemble

Ordinary Metropolis and Monte Carlo methods are designed to sample from distributions that are functions such as the Boltzmann distribution. When a distribution is not a function such as the Dirac-delta (which is the limit of a sequence of functions but not itself a function) used for the microcanonical ensemble, a computer cannot calculate it directly. In those cases the distribution must be modeled in another way. There are two main ways of modeling microcanonical ensembles: the Demon algorithm which is an MCMC algorithm different from Metropolis and the ODE-based Hamiltonian flow method.

8.3.1 Demon Algorithm

In 1871, James Clerk Maxwell conceived of a thought experiment that challenged the idea of the laws of thermodynamics as fixed physical laws. In this thought experiment, he posited a "demon" that could affect the state of a container of molecules so as to violate the second law of thermodynamics [101]:

> One of the best established facts in thermodynamics is that it is impossible in a system enclosed in an envelope which permits neither change of volume nor passage of heat, and in which both the temperature and the pressure are everywhere the same, to produce any inequality of temperature or of pressure without the expenditure of work. This is the second law of thermodynamics, and it is undoubtedly true as long as we can deal with bodies only in mass, and have no power of perceiving or handling the separate molecules of which they are made up. But if we conceive a being whose faculties are so sharpened that he can follow every molecule in its course, such a being, whose attributes are still as essentially finite as our own, would be able to do what is at present impossible to us. For we have seen that the molecules in a vessel full of air at uniform temperature are moving with velocities by no means uniform, though the mean velocity of any great number of them, arbitrarily selected, is almost exactly uniform. Now let us suppose that such a vessel is divided into two portions, A and B, by a division in which there is a small hole, and that a being, who can see the individual molecules, opens and closes this hole, so as to allow only the swifter molecules to pass from A to B, and only the swifter molecules to pass from A to B, and only the slower ones to pass from B to A. He will thus, without expenditure of work, raise the temperature of B and lower that of A, in contradiction to the second law of thermodynamics.

Maxwell's arguments are controversial. One of the main objections is that the "being" which some call Maxwell's Demon is also a part of the system and, as the entropy of the vessel decreases, the demon's entropy must increase. From a quantum point of view, the decrease of entropy in the system corresponds to a decrease in information content, i.e., the demon is deleting the information contained in the vessel and must, therefore, be taking that information into itself somehow for information cannot be lost. This controversy is very much alive today in discussions of information loss in black holes and the holographic principle [54, 122]. In that case, the black hole acts as the demon, swallowing up information from the universe. By the holographic principle (such as the anti-de Sitter/conformal field theory correspondence), however, this information must somehow be preserved on the surface of the black hole event horizon as the black hole grows and its event horizon grows with it [15]. Or alternatively quantum fluctuations in the horizon allow the information to escape. Thus, the hypothesis is that the universe will not permit Maxwell's demon to exist unless it transfers every piece of lost entropy onto itself, and the second law remains absolute. This idea is still the subject of considerable research, however, and is far from settled.

8.3 Microcanonical Ensemble

The demon algorithm was developed for microcanonical simulations by David Creutz. Creutz's algorithm takes Maxwell's demon and turns it around so that, rather than decreasing entropy, it increases it to maximum [39]. The purpose of the demon algorithm is to simulate microcanonical ensembles by injecting an extra degree of freedom into a system known as a demon. This demon essentially absorbs and releases small amounts of energy from and into the ensemble allowing the ensemble to change its state while maintaining a fixed energy.

First, let us consider how we would do a Markov Chain Monte Carlo simulation of a molecular system without a demon. Because the ensemble has a Dirac-delta function for a distribution,

$$\delta(E - H_N)$$

the system must find a way to move from state s to state s' such that $H_N^s = H_N^{s'} = E$. The reason is that states where $H_N^s \neq E$ have zero probability of occurring so we need to avoid them completely. Clearly, the starting state must be chosen to have energy E. In a non-interacting system, the solution to how to obtain a new state from there is simple. "Trade" energy between particles, i.e., select one molecule, i, to lose energy, ΔE, and one molecule, j, to gain an equal amount of energy. Doing Monte Carlo on a non-interacting system, however, is usually not very interesting. We want to study particles that communicate with each other.

Now, consider an interacting system. Here, the problem is that changing the state of one particle changes the energy of all the other particles. (For short-range interacting systems like Ising models, the problem is simpler.) For example, if we have a system of 2D point vortices and move one vortex, then it changes the interacting energy with all the other ones. We need to move another vortex, therefore, to correct for it, but changing another one also affects all the other vortices. There is a domino effect in trying to move particles in an interacting system while keeping the energy constant. We can, however, find moves that do this (typically involving three vortices), but, even so, this does not solve the second problem with fixed energy MCMC which is that we no longer have a way to measure the temperature. When simulating the Boltzmann distribution the temperature is fixed while the energy fluctuates, but since we have a formula for the energy (and need to calculate it at every step anyway) we always know both the energy and the temperature. In a fixed energy simulation we always know the energy, but we do not always have a neat formula for the temperature. The temperature itself is important, especially if a system has negative temperature states or something particularly interesting is supposed to happen at a particular temperature (like a phase transition), but it also is useful for calculating the specific heat as the relationship between energy and temperature. As we saw in the chapter on plasmas negative specific heat is an indicator of metastable states, an important consideration in fixed energy systems. Thus, we would like a way to measure the temperature. One way is to measure the kinetic energy per particle and use the equipartition theorem to relate that to the temperature. But what if there is no kinetic energy? For example, what if it is "infinite" as in 2D vortices?

Creutz's algorithm provides solutions to both the problems mentioned: (1) how to keep the energy constant in an interacting system and (2) how to measure the temperature in a fixed energy system. The concept is simple. A variable called the demon acts as a repository for energy. When the system is initialized, the total energy of the system plus demon equals the desired fixed energy. Monte Carlo moves are performed as they normally would in the Metropolis algorithm, however, at the acceptance/rejection stage the move is accepted only if the demon contains enough energy to cover any increase in the energy of the system. For example, suppose the energy must remain fixed at 100 units. If the demon has 12 units of energy, then state s must have 88 units. If state s' requires 104 units, then the move is rejected. If it requires, say, 96, then the move is accepted and 8 units of energy are subtracted from the demon so the new configuration has the *demon* with 4 units. Decreases in energy are always accepted as in standard Metropolis. Typically, the average demon energy is considerably less than the energy of the system, so that its influence is minimal. Moreover, Creutz showed that, because it is a single degree of freedom, the average energy of the demon is equal to the temperature of the system. In this respect, the demon acts as a thermometer, introducing a single, negligible degree of freedom to the system as well as measuring the temperature of the system.

It is clear that Creutz's demon is not playing the same role as Maxwell's. Rather than reducing the entropy of the system, Creutz's demon enables the system to reach its maximum entropy without requiring special, energy conserving moves. It is, however, an artificial device, and, in some cases, such as at very high temperatures, the demon can have a non-negligible effect on the behavior of the system. It is also useless as-is for negative temperature states. Another problem arises when the cost of an acceptance/rejection algorithm is too high (such as quantum Monte Carlo simulations of fermions), and we would like every new state to be an "accepted" state rather than rejecting half of the states. In these cases, we turn to Hamiltonian flow algorithms.

8.3.2 Hamiltonian Flow

Hamiltonian flow algorithms are not, as they may seem from the name, time-evolution simulations in any realistic sense, although they can be. The purpose of the flow algorithm is to use the principles of Hamiltonian mechanics to explore the fixed energy "shell" of an ensemble without resorting to acceptance rejection schemes or demons. Any Hamiltonian system can be subject to a Hamiltonian flow algorithm, including vortex systems.

The algorithm works as follows [27, 28]: the system is initialized to the desired energy, E. Then the set of Hamiltonian equations with generic canonical coordinates $q_1, \ldots, q_N$ and $p_1, \ldots, p_N$,

$$\frac{dp_i}{dt} = -\frac{\partial H_N}{\partial q_i}, \quad \frac{dq_i}{dt} = \frac{\partial H_N}{\partial p_i},$$

are solved so that the state at some time in the future is found. It does not matter how far into the future (or past for that matter) we solve and we do not need to solve in time order. The key is that the Hamiltonian ODEs conserve the energy of the system. Each solution to the evolution equations acts like a move in MCMC but every move is accepted. Thus, the statistical properties of the system can be found by averaging over the different time evolutions. In particular, states far in time from the initial state are most reliable because the initial conditions will likely have been "forgotten" in a statistical sense. If the system has a stable equilibrium, it will reach it at a time far from its initial state. If it does not have a stable equilibrium, e.g., it is a chaotic system, then equilibrium statistical mechanics is not applicable because its basic requirement is that an equilibrium exists.

The drawback of Hamiltonian flow is that it requires solving a potentially large number of ODEs. This is much more computationally intensive than making random moves. Therefore, it is best used in cases where a demon algorithm is undesirable.

8.4 Numerical Simulations in Vorticity

Numerical simulations have been applied to vortex flows since at least the 1970s [36]. There are several seminal numerical simulations in vortex statistics and dynamics. The emergence of vortex filaments and tangles, for example, from turbulence simulations was shown by She et al. [136]. Recent trends in topological fluid mechanics (essentially the study of fluids as topologies of vortices and other defect structures) have applied computing to knot and braid theory in fluid mechanics [128]. Tangles are also an important area of numerical simulation [16]. Numerical PDEs have revealed many interesting vortex structures in quantum superfluids [40, 41] and classical fluids, such as merger [72] and reconnection [23]. High performance computing has allowed billion vortex particle models to take on very complex airplane wakes by modeling vorticity as if it were made up of small particles [31]. It is likely that vortex research will continue in this direction and increasingly be accepted into applications and applied research.

8.5 Concluding Remarks

This chapter has given you a whirlwind tour of the methods that are used in statistical physics (and quantum physics). In the next chapter, we will see one of these methods, the canonical MCMC algorithm, put into action to demonstrate an analytical result we saw in the previous chapter but applied to deep ocean convection rather than plasmas.

Chapter 9
Quasi-2D Monte Carlo in Deep Ocean Convection

In this chapter, we present some results from our papers [7, 8]; these results bring together hydrodynamics, statistics, and MCMC in a complete illustration of the solution to a research problem using both analysis and computation.

Deep ocean convection due to localized surface cooling is an important phenomenon that has been extensively studied in field observations (e.g., Greenland Sea), laboratory experiments, numerical, and theoretical models. In lab experiments on convective turbulence in homogeneous rotating fluids, a transition is observed from 3D turbulence to quasi-2D rotationally controlled vortex structures with axes parallel to the rotational axis [102, 124]. When a large number of quasi-2D vortex structures are present, they can appear in arrays or lattices, and these have been commonly studied as fully 2D arrays, in one layer [70, 88, 117], or in two layers as a heton model [43, 85]. However, it is an open question how quasi-2D structures nearer to the transition and/or with small inter-vortex distance behave because 2D models are not adequate to describe them. Of considerable interest is the relative cross-sectional size of these arrays—especially in terms of inter-vortex distance and curvature—that results from conservation of angular momentum [96].

The trouble with the 2D model is that 3D effects become exceptionally important when rotation is weak or counter-currents are strong, and so extensions have been made to introduce three-dimensionality without losing the advantages of the 2D logarithmic interaction and without resorting to complete 3D turbulence modeling. The equilibrium statistical model of [43] and dynamical study of [85] are two such attempts each layering two 2D Point Vortex Gases one on top of the other and introducing interaction between layers giving a pseudo-3D flavor. The heton model also includes f-plane planetary rotational effects (i.e., the Coriolis force) in the form of a Rossby radius of deformation. This slightly more complicated model has allowed for some interesting results in determining the statistical distribution of vortices in the layers from low-interaction levels where the distribution is essentially normal to high-interaction where it is essentially uniform with a sharp cut-off at the boundary [12].

An extension of the two layered approach is a multi-layered approach, and, taking this approach to its logical conclusion, we arrive at what one could term the "infinite"-layered model but what is more commonly called the nearly parallel vortex filament model [78, 92]. This model has much more structure than the heton model but neglects planetary effects. This is the model studied in this chapter; therefore, we devote Sect. 9.1 to explaining it more completely. To outline it briefly here, it is a model in which the 3D vorticity field is represented as a large number, N, of vortex filaments. (Pictured in Fig. 9.1.) Each vortex filament, j, is a curve in space that we represent as a complex function, $\psi_j(\sigma) \in \mathbb{C}$, where $\sigma \in \mathbb{R}$ is a parameter, and each one has a certain strength λ_j. The vortex curves are all nearly parallel (in an asymptotic sense) to the z-axis, hence, *nearly parallel* vortex filaments. Because they are quasi-2D, they have a 2D interaction between points in the same plane on different filaments. Stretching is minimal. Internal fluctuations are represented with a local-induction approximation (LIA), explained in Sect. 9.1. The best benefit of the system is that it is finite Hamiltonian and fits into the same distribution as the original Onsager model. Since our investigation into 3D effects are based on theoretical analysis of the probability distribution and Monte Carlo simulations, this simplicity is crucial.

In our analysis of the equilibrium statistics of this system, we focus on the most critical statistic: size, defined as the second moment of the statistical distribution (Sect. 9.1). Size is important in the study of ocean convection because, of all statistical behaviors, it is the most visible and the most measurable. Clearly, the third dimensional variations in filaments have some effect on overall system size, but it is an open question whether there is a phase transition due to increasing 3D effects in a quasi-2D system and whether the local self-induced variations act as a counter to the expansive effect of interaction potential or challenge the squeezing effect of conservation of angular momentum. To determine this, we set our goal to achieving an explicit, closed-form approximation for the system size and confirming its accuracy computationally.

In order to determine the size of the system analytically, we use a mean-field approach, described in Sect. 9.2, in which the system of N vortex filament structures is replaced with two vortex filaments, an ordinary one a mean distance from the origin with strength 1 and a perfectly straight one at the origin containing the mean center of vorticity (having a strength of $N-1$). We use the statistics of the outer vortex to approximate the behavior of any given vortex in the system. Our approximation is a special case of the rigorous mean-field approach of [92] and is justified in its existence. To justify its accuracy, we turn to Monte Carlo simulations of the original system, described in Sect. 9.4.

Our results indicate that not only is the mean-field approximation accurate but that the size of the system experiences a significant transition in the parameter β. We find that a β_0 exists such that the change in the size with respect to β switches direction. We are able to calculate an explicit, closed-form formula for the squared size of the system, R^2, and confirm the formula with Monte Carlo measurements (Sect. 9.4).

9.1 The Nearly Parallel Vortex Filament Model's Entropy-Driven Shift

9.1.1 Background

The full boundary value problem of deep ocean convection-driven vortex structures is exceedingly complicated and not necessarily useful. To extract general physical principles the full complexity is not required. Rather, the Onsager model and its related models [12, 117] simplify the 2D Euler problem to the bare minimum required for a meaningful statistical mechanical approach: no boundaries, discrete vortex structures with no individual cross-sections (points), and a large number of conserved quantities such as energy, angular momentum, vorticity, etc. Leaving out 3D effects, these models fail in cases where inter-vortex distance is small relative to the distance a filament's curve travels in the plane. (Filaments are able to cross each other due to the viscosity in their cores, which is not represented explicitly in inviscid models.) The nearly parallel vortex filament model adds a small element of true three-dimensionality to the 2D Point-Vortex Gas, enough to explore what happens when plane-position variations along the filament begin to dominate vortex–vortex interaction and angular momentum.

Without describing the mathematics in detail yet, 3D vortex filaments behave much like springs, and stretching the filament increases its energy. In systems of nearly parallel vortex filaments, there are two kinds of energy: self-energy that increases with localized stretching and interaction energy that increases as vortex structures come closer together. The other component, angular momentum, is conserved but unaffected by temperature. Because the convective-rotation is nearly a rigid rotation, angular momentum increases with the square of distance from the axis of rotation (the z-axis in our case). When the self-energy is insignificant (or zero), the interaction energy and angular momentum compete and determine the system's size.

For the case of zero self-energy (and zero entropy) [88] give a formula for the system size,

$$R^2 = \frac{\Lambda \beta}{4\mu}, \tag{9.1}$$

where

$$R^2 = \lim_{N \to \infty} \int ds N^{-1} \sum_{j=1}^{N} |z_j|^2 p(s) \tag{9.2}$$

is the second-moment of $p(s) = P_N^{2D}(s)$ in the infinite-N limit with the necessary redefinition of inverse temperature, $\beta' = \beta N$ where β' is kept constant (called a *non-extensive thermodynamic limit*), and Λ is the total vortex strength, also kept constant in N.

9.1.2 Hypothesis

Our hypothesis is that when filament variations become large compared to inter-vortex distance the following process becomes dominant: when a vortex moves away from the center, potential energy decreases and angular momentum increases. In a 2D model point vortex model consisting of like-signed charges, this would cause other vortices to move inward to restore the balance (Process 1). In a 3D model, this can also happen, but, alternatively, part of the *same filament* can move closer to the center, leaving the other filaments fixed (Process 2). This increase in variation restores the balance but increases the self-energy, so the change toward the center is not as significant as it would be if a different filament moved to compensate. The overall effect of Process 2 is expansion. Process 1 results in a lower total energy than Process 2 and so, at low positive temperatures, this is the dominating process, but, at high positive temperatures, Process 2 dominates because Process 1 does not increase the entropy of the system, while Process 2 does. Thus, the shift from Process 1 to Process 2 is entropy driven.

As entropy becomes more significant, this expansive effect begins to dominate. We hypothesize that as the cost of increasing self-energy and interaction energy (the total energy) decreases with increasing temperature, a double effect occurs: the angular momentum contracts the system, decreasing inter-vortex distance at first, but the decreased distance causes the variations to become more significant without making the filaments any less straight. The expansion effect begins to dominate the angular momentum's contraction effect, eventually stopping and reversing the contraction of the system's size. Showing that this occurs would effectively validate the hypothesis. However, we go one step further and give an explicit formula for R^2, which allows us to explore the parameter space as completely as possible.

9.1.3 Mathematical Model

With our hypothesis given, the following is an overview of the mathematical model we employ: This quasi-2D model [78] is derived rigorously from the Navier–Stokes equations and represents vorticity as a bundle of N filaments that are nearly parallel to the z-axis. The model has a Hamiltonian,

$$H_N = \alpha \int_0^L d\sigma \sum_{k=1}^N \frac{1}{2} \left| \frac{\partial \psi_k(\sigma)}{\partial \sigma} \right|^2 - \int_0^L d\sigma \sum_{k=1}^N \sum_{i>k}^N \log|\psi_i(\sigma) - \psi_k(\sigma)|, \quad (9.3)$$

where $\psi_j(\sigma) = x_j(\sigma) + i y_j(\sigma)$ is the position of vortex j at position σ along its length, the circulation constant is same for all vortices and set to 1, and α is the core structure constant [78]. The position in the complex plane, $\psi_j(\sigma)$, is assumed to be periodic in σ with period L. The angular momentum is

9.2 A Mean-Field Approach

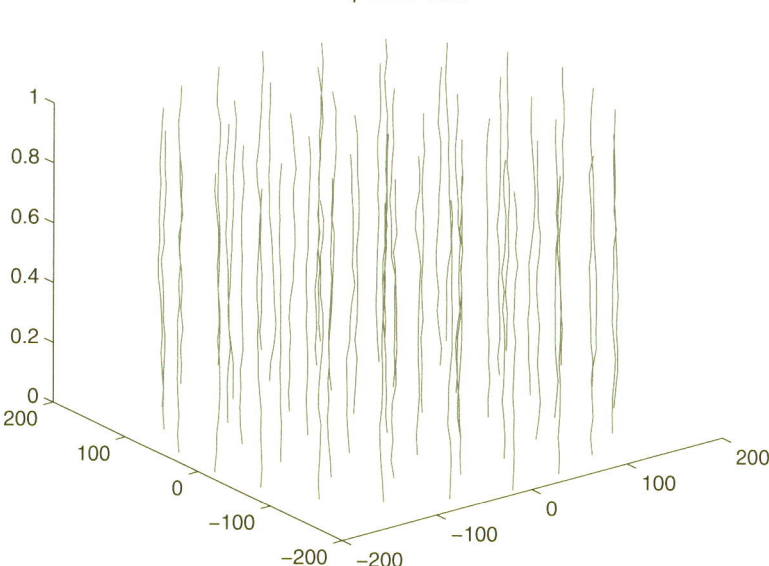

Fig. 9.1: This output from our Monte Carlo simulation of a single sample illustrates how the moderate-temperature nearly parallel vortex filament model appears. In a high-temperature case, filaments may cross while in low-temperature cases the filaments' variations are too small to be visible

$$I_N = \sum_i^N \int_0^L d\sigma \, |\psi_i(\sigma)|^2. \tag{9.4}$$

The Gibbs distribution for this system, P_N, has the same form as P_N^{2D}:

$$P_N(s) = \frac{1}{Z_N} e^{-\beta H_N - \mu I_N}, \tag{9.5}$$

where $Z_N = \sum_s e^{-\beta H_N - \mu I_N}$.

9.2 A Mean-Field Approach

The first step in any analysis of a statistical mechanical system is to calculate or approximate the normalizing factor, Z_N, called the *partition function*. The partition function for the quasi-2D system is

$$Z_N = \int D\psi_1 \cdots \int D\psi_N \exp(S_N), \tag{9.6}$$

where $D\psi_i$ represents functional integration over all paths for each filament i (also known as a Feynman or Feynman–Kac integral, [48]). The functional $S_N = -\beta H_N - \mu I_N$ is the *action*.

We have the most-probable free energy from the following formula [134]:

$$F = -\frac{1}{\beta} \log Z_N. \qquad (9.7)$$

As is well known, the state that gives the minimum free energy is the most-probable state of the system. In this case the state consists of the positions of the filaments $\{\psi_i\}_{i=1...N}$. The interaction term, second term in (9.3), makes a direct analytical solution impossible with current knowledge. While the other terms in S_N, the self-induction, from the first term in (9.3), and the conservation of angular momentum term, $-\mu I_N$, are negative definite quadratic and yield a normally distributed Gibbs distribution that we can functionally integrate, the logarithmic term must be approximated. The simplest way to do the approximation is a mean-field theory which will reduce the problem from N coupled (interacting) filaments to N uncoupled (non-interacting) filaments.

Although [92] have made such an approximation and rigorously derived a mean-field evolution PDE for the probability distribution of the vortices in the complex plane, their PDE takes the form of a nonlinear Schrödinger equation that is not analytically solvable (even in equilibrium), again because of the interaction term. Our mean-field theory is a special case of theirs.

A reasonable approach to a mean-field theory is to change the interaction between each filament and all the other filaments to an interaction between a single filament and one perfectly straight filament at the origin with the combined strength of all the filaments. Each pair of filaments i and j have a square distance associated with each plane σ: $|\psi_i(\sigma) - \psi_j(\sigma)|^2$. If we combine all the filaments into one average filament, it will be at the origin. Therefore, the mean square distance between a filament and any other filament can be approximated by the square distance between that filament and the origin. If we take $\langle \cdot \rangle$ to mean average, then $\langle |\psi_i(\sigma) - \psi_j(\sigma)|^2 \rangle \approx |\psi_i(\sigma)|^2$ where the average is over filaments j. Furthermore, assuming filaments are placed uniformly, the mean square distance of a filament from the origin, $\langle |\psi_i|^2 \rangle = N^{-1} \sum_{i=1}^{N} \int_0^L d\sigma |\psi_i(\sigma)|^2 = N^{-1} I_N$. Given these assumptions the interaction takes the following form:

$$\int_0^L d\sigma \frac{1}{4} \sum_{i=1}^{N} \sum_{j=1}^{N} \log |\psi_i(\sigma) - \psi_j(\sigma)|^2 = \frac{N^2}{4} \log \frac{I_N}{N}. \qquad (9.8)$$

Of course, these assumptions remain to be validated.

We can take the mean-field action to be

$$S_N^{mf} = \frac{LN^2 \beta}{4} \log \frac{I_N}{N} - \int_0^L d\sigma \left[\sum_{k=1}^{N} \frac{\beta \alpha}{2} \left| \frac{\partial \psi_k(\sigma)}{\partial \sigma} \right|^2 + \mu \sum_{k=1}^{N} |\psi_k(\sigma)|^2 \right], \qquad (9.9)$$

where the first term is now mean-field and the other two are the same as before.

9.3 Solving for the Square Radius

Before we begin to calculate the partition function, we must deal with another problem: we still cannot integrate (9.6) using this action because the interaction term is still a function of ψ, so we make another approximation, adding a spherical constraint [20, 64] on the angular momentum,

$$\delta\left(\int_0^L d\sigma \left[I_N - NR^2\right]\right), \qquad (9.10)$$

which has integral representation,

$$\int_{-\infty+i\tau_0}^{\infty+i\tau_0} \frac{d\tau}{2\pi} \exp \int_0^L -i\tau \left[I_N - NR^2\right], \qquad (9.11)$$

where R^2 is defined by (9.2). The spherical-mean-field partition function is now

$$Z_N^{smf} = \int D\psi_1 \cdots \int D\psi_N \exp\left(S_N^{mf}\right) \int_{-\infty}^{\infty} \frac{d\tau}{2\pi} \exp \int_0^L d\sigma - i\tau \left[I_N - NR^2\right]. \quad (9.12)$$

The spherical constraint does not alter the statistics of the system significantly because the angular momentum already has an implicit preferred value, NR^2, we are simply making it explicit.

9.3 Solving for the Square Radius

We present here two approaches to deriving an analytical solution for R^2. The first is based on a quantum harmonic oscillator in a continuum and the second uses a spherical model on the discrete model.

9.3.1 Harmonic Oscillator Approach

We now solve Z_N^{smf} in closed-form in the limit as $N \to \infty$: Since the exponents are all negative definite, we can interchange the integrals and combine exponents:

$$Z_N^{smf} = \int \frac{d\tau}{2\pi} \int D\psi_1 \cdots \int D\psi_N \exp\left(S_N^{smf}\right), \qquad (9.13)$$

where the combined action functional is $S_N^{smf} = \sum_{k=1}^N S_k$, and the single filament action is

$$S_k = \left[\frac{\beta LN \log(R^2)}{4} - \frac{1}{2}\int_0^L d\sigma\, \alpha\beta \left|\frac{\partial \psi_k(\sigma)}{\partial \sigma}\right|^2 + (i\tau + 2\mu)|\psi_k(\sigma)|^2 - iR^2\tau\right].$$

Because $\{\psi_k\}$ are statistically independent for all k, we can drop the k subscript. Therefore, the total action is simply a multiple of the single filament action: $S_N^{smf} = NS$, which makes the partition function

$$Z_N^{smf} = \int \frac{d\tau}{2\pi} \left\{ \int D\psi \exp S \right\}^N. \tag{9.14}$$

Now we define the *non-dimensional free energy*, $f[i\tau] = \beta F$. Using the formula in (9.7),

$$f[i\tau] = -\log\left[\int D\psi \exp(S)\right] \tag{9.15}$$

where

$$S = \left[\beta'\frac{L\log(R^2)}{4} - \frac{1}{2}\int_0^L d\sigma\, \alpha'\beta' |\frac{\partial \psi(\sigma)}{\partial \sigma}|^2 + (i\tau + 2\mu)|\psi(\sigma)|^2 - iR^2\tau\right], \tag{9.16}$$

$\beta' = \beta N$, and $\alpha' = \alpha N^{-1}$ and

$$Z_N^{smf} = \int_{-\infty}^{\infty} \frac{d\tau}{2\pi} \exp(-Nf[i\tau]). \tag{9.17}$$

We now have a partition function we can solve with steepest-descent methods if we find an expression for f. The functional $f[i\tau]$ is the energy of a 2-D quantum harmonic oscillator with a constant force and simply evaluated with Green's function methods [26].

The free-energy, (9.15), involves a simple harmonic oscillator with a constant external force, and we can re-write it,

$$f[i\tau] = -\frac{1}{2}i\tau LR^2 - \frac{\beta' L\log(R^2)}{4} - \ln h[i\tau]. \tag{9.18}$$

Here h is the partition function for a quantum harmonic oscillator in imaginary time,

$$h[i\tau] = \int D\psi \exp\left(\int_0^L d\sigma -\frac{1}{2}m[|\partial_\sigma \psi|^2 + \omega^2|\psi|^2]\right), \tag{9.19}$$

which has the well-known solution for periodic paths in (2+1)-D where we have integrated the end-points over the whole plane as well,

$$h[i\tau] = \frac{e^{-\omega L}}{(e^{-\omega L} - 1)^2}, \tag{9.20}$$

where $m = \alpha'\beta'$ and $\omega^2 = (i\tau + 2\mu)/(\alpha'\beta')$ [26, 152].

Let us make a change of variables $\lambda = i\tau + 2\mu$. Then the free-energy reads

$$f[\lambda] = (\mu - \frac{1}{2}\lambda)LR^2 - \frac{\beta' L\log(R^2)}{4} - \ln \frac{e^{-\omega L}}{(e^{-\omega L} - 1)^2}, \tag{9.21}$$

where $\omega = \sqrt{\lambda/(\alpha'\beta')}$.

9.3 Solving for the Square Radius

Now that we have a formula for f we can apply the saddle point or steepest-descent method. (For discussion of this method see Sect. 9.3.2 as well as the original paper of Berlin and Kac [20].) The intuition is that, as $N \to \infty$ in the partition function, only the minimum energy will contribute to the integral, i.e., at infinite N, the exponential behaves like a Dirac-delta function, so

$$f_\infty = \lim_{N \to \infty} -\frac{1}{N} \ln Z_N^{smf} = f[\eta], \tag{9.22}$$

where η is such that $\partial f[\lambda]/\partial \lambda|_\eta = 0$ [20, 64].

First we can make a simplification by ridding (9.21) of R^2. We know that R^2 will minimize f and so $\partial f/\partial R^2 = 0$. Therefore,

$$R^2 = \frac{\beta'}{4(\mu - \lambda/2)}. \tag{9.23}$$

Substituting the left side of (9.23) for R^2 in (9.21), we get

$$f[\lambda] = \frac{\beta' L}{4} + \sqrt{\frac{\lambda}{\alpha'\beta'}} L + 2\log \left| \exp\left(-\sqrt{\frac{\lambda}{\alpha'\beta'}} L\right) - 1 \right|$$
$$- \frac{\beta' L}{4} \log \frac{\beta'}{4(\mu - \lambda/2)}. \tag{9.24}$$

We could take the derivative of (9.24) and set it equal to zero to obtain η. However, doing so yields a transcendental equation that needs to be solved numerically. Since our goal is to obtain an explicit formula, we choose to study the system as $L \to \infty$. In fact such an approach is justified by the assumptions of the model that L have larger order than the rest of the system's dimensions. (If this were a quantum system, this procedure would be equivalent to finding the energy of the ground state. Hence, we call this energy f_{grnd}.) Taking the limit on (9.24) yields the free energy per unit length in which η can be solved for

$$f_{\text{grnd}}[\eta] = \frac{\beta'}{4} + \sqrt{\eta/(\alpha'\beta')} - \frac{\beta'}{4} \log\left(\frac{\beta'}{4(\mu - \eta/2)}\right), \tag{9.25}$$

where

$$\eta = 2\mu - \frac{1}{8}\beta'(-\beta'^2\alpha' + \sqrt{\beta'^4\alpha'^2 + 32\alpha'\beta'\mu}) \tag{9.26}$$

gives physical results.

With η explicit, we can give a full formula for R^2,

$$R^2 = \frac{\beta'^2\alpha' + \sqrt{\beta'^4\alpha'^2 + 32\alpha'\beta'\mu}}{8\alpha'\beta'\mu}. \tag{9.27}$$

Through several approximations, we have obtained an explicit formula for the free energy of the system and R^2.

9.3.2 Spherical Method Approach

The spherical model was first proposed in a seminal paper of Berlin and Kac [20], in which they were able to solve for the partition function of an Ising model given that the site spins satisfied a spherical constraint, meaning that the squares of the spins all added up to a fixed number. The method relies on what is known as the saddle point or steepest-descent approximation method which is exact only for an infinite number of lattice sites.

In general the steepest-descent or saddle-point approximation applies to integrals of the form

$$\int_a^b dx\, e^{-Nf(x)}, \qquad (9.28)$$

where $f(x)$ is a twice-differentiable function, N is large, and a and b may be infinite. A special case, called Laplace's method, concerns real-valued $f(x)$ with a finite minimum value.

The intuition is that if x_0 is a point such that $f(x_0) < f(x) \forall x \neq x_0$, i.e., it is a global minimum, then, if we multiply $f(x_0)$ by a number N, $Nf(x) - Nf(x_0)$ will be larger than just $f(x) - f(x_0)$ for any $x \neq x_0$. If $N \to \infty$, then the gap is infinite. For such large N, the only significant contribution to the integral comes from the value of the integrand at x_0. Therefore,

$$\lim_{N \to \infty} \left[\int_a^b dx\, e^{-Nf(x)} \right]^{1/N} = e^{-f(x_0)}, \qquad (9.29)$$

or

$$\lim_{N \to \infty} -\frac{1}{N} \log \int_a^b dx\, e^{-Nf(x)} = f(x_0) \qquad (9.30)$$

[20, 64]. A proof is easily obtained using a Taylor expansion of $f(x)$ about x_0 to quadratic degree.

The theorem follows from a steepest-descent argument [20, 139]. In employing steepest descent, we need to start with an integral of the form, $\int dx\, e^{-MF[x]}$. Then if $F[x_0] < F[x] \forall x \neq x_0$, i.e., its minimum value is at x_0,

$$\lim_{M \to \infty} M^{-1} \log \left(\int dx\, e^{-MF[x]} \right) = F[x_0].$$

This works because, as M increases, the distribution that $e^{-MF[x]}$ represents becomes narrower and focuses on x_0 until the distribution has zero value at all other x.

Let us assume that N is large and take the limit over M first. We can pull the constant terms out of the integral of $Z'_N(M)$ over Ψ. Because F is positive definite, under Fubini's theorem we may switch the integrals to obtain

$$Z'_N(M) = e^{N^2 \beta L/4 \log R^2 - \mu NLR^2} \int_{-\infty}^{\infty} \frac{d\sigma}{2\pi} \int d\Psi\, e^{-NF'[i\sigma]}, \qquad (9.31)$$

9.3 Solving for the Square Radius

where

$$F'[i\sigma] = \alpha\beta \sum_{k=1}^{M} \frac{1}{2} \frac{|\psi(k+1) - \psi(k)|^2}{\varepsilon} + i\sigma \left(\sum_{k=1}^{M} |\psi(k)|^2 - MR^2 \right), \quad (9.32)$$

is the part of the free energy still dependent on Ψ.

The interior integral needs evaluation. Let $s = i\sigma$ and define a new partition function,

$$Z'(M) = \int_{-i\infty}^{i\infty} \frac{ds}{2\pi} \int d\Psi \, e^{-F'[s]}, \quad (9.33)$$

be that interior. Let us put F' in matrix form:

$$F'[s] = -sMR^2 + K\Psi^\dagger A\Psi + s\Psi^\dagger \Psi, \quad (9.34)$$

where $K = \alpha\beta/\varepsilon$, and the $M \times M$ matrix A has the form

$$A_{i,i} = 1, \quad (9.35)$$
$$A_{i,i+1} = A_{i+1,i} = A_{1,M} = A_{M,1} = -\tfrac{1}{2}, \quad (9.36)$$
$$A_{i,j} = 0 \quad \text{other} \quad i,j. \quad (9.37)$$

The integral in Eq. (9.33) is Gaussian. We can evaluate it, knowing the eigenvalues of the matrix A. These eigenvalues have the form $\lambda_i = 1 - \cos(2\pi(i-1)/M)$, (not related to the previous use of λ_i as strength of vorticity) [20, 92] and so

$$Z'(M) = \int_{-i\infty}^{i\infty} \frac{ds}{2\pi} e^{sMR^2} \pi^M \prod_i \frac{1}{s + K\lambda_i}. \quad (9.38)$$

We need to put $Z'(M)$ back into the correct form for steepest descent. Following the example of Berlin and Kac [20] and standard asymptotic analysis [139], let $s = K(\eta - 1)$ then

$$Z'(M) = \int_{-i\infty}^{i\infty} \frac{d\eta}{2\pi} K\pi^M (\eta - 1)^{-1} e^{-M\log(K)} e^{Mf[\eta]}, \quad (9.39)$$

where

$$f[\eta] = KR^2(\eta - 1) - M^{-1} \sum_{i=2}^{M} \log(\eta - \cos(2\pi(i-1)/M)), \quad (9.40)$$

and $\eta \geq 1$.

We leave the $i = 1$ term out of the sum in $f[\eta]$ so that we can evaluate f further by taking the limit on the second term,

$$\lim_{M \to \infty} M^{-1} \sum_{i=2}^{M} \log(\eta - \cos(2\pi(i-1)/M)) = \frac{1}{2\pi} \int_0^{2\pi} d\omega \, \log(\eta - \cos(\omega)),$$

which gives

$$f[\eta] = KR^2(\eta - 1) - \frac{1}{2\pi}\int_0^{2\pi} d\omega \log(\eta - \cos(\omega))$$
$$= KR^2(\eta - 1) - \log(\eta + (\eta^2 - 1)^{\frac{1}{2}}). \quad (9.41)$$

To apply steepest descent, we determine the saddle point $\eta = \eta_0$ where $f[\eta]$ has its minimum value, $f[\eta_0]$. Taking the derivative and setting it equal to zero,

$$\frac{\partial f}{\partial \eta} = KR^2 - \frac{1}{\sqrt{\eta^2 - 1}} = 0, \quad (9.42)$$

implies

$$\eta_0 = \sqrt{\frac{1}{(KR^2)^2} + 1}. \quad (9.43)$$

Having evaluated f, we can give an equation for the original free energy. Letting $N\beta \to \beta'$ and $\alpha/N \to \alpha'$,

$$F[\eta_0] = -\beta' L/4 \log R^2 + \mu L R^2 + M \log(K) - M f[\eta_0], \quad (9.44)$$

and evaluate it as $M \to \infty$. We will drop primes on β and α now.

The term in the limit $M \log(K)$ does not depend on R^2, and it is an unnecessary component representing the entropy of the broken filaments in the non-interacting case, and we drop it. Now we fill in the expressions for η_0 and K as defined above:

$$\lim_{M\to\infty} M f[\eta_0] = \lim_{M\to\infty} KR^2(\eta_0 - 1) - M\log(\eta_0 + (\eta_0^2 - 1)^{\frac{1}{2}})$$

$$= \lim_{M\to\infty} MKR^2\left(\sqrt{\frac{1}{(KR^2)^2} + 1} - 1\right)$$

$$- M\log\left(\sqrt{\frac{1}{(KR^2)^2} + 1} + \frac{1}{KR^2}\right)$$

$$= \lim_{M\to\infty} M\frac{\alpha\beta M}{L}R^2\left(\sqrt{\frac{1}{(\frac{\alpha\beta M}{L}R^2)^2} + 1} - 1\right)$$

$$- M\log\left(\sqrt{\frac{1}{(\frac{\alpha\beta M}{L}R^2)^2} + 1} + \frac{1}{\frac{\alpha\beta M}{L}R^2}\right)$$

which, because it is an energy for a filament, ought to be finite. The first term is the energy of the filament, E_{fil}, and the second, the entropy, S_{fil}, and each by itself is finite, so we take each limit separately.

The energy:

$$E_{\text{fil}} = \lim_{M\to\infty} M\frac{\alpha\beta M}{L}R^2\left(\sqrt{\frac{L^2}{(\alpha\beta MR^2)^2} + 1} - 1\right) = \frac{L}{2\alpha\beta R^2} \quad (9.45)$$

The entropy:

$$S_{\text{fil}} = \lim_{M \to \infty} M \log \left(\sqrt{\frac{1}{(\frac{\alpha \beta M}{L} R^2)^2} + 1} + \frac{1}{\frac{\alpha \beta M}{L} R^2} \right) = \frac{L}{\alpha \beta R^2} \quad (9.46)$$

These two results imply that

$$\lim_{M \to \infty} M f[\eta_0] = -\frac{L}{2\alpha\beta R^2}$$

and

$$F[\eta_0] = L\mu R^2 - \beta L/4 \log R^2 + \frac{L}{2\alpha\beta R^2}. \quad (9.47)$$

We minimize with respect to R^2,

$$\frac{\partial F}{\partial R^2} = L\mu - \frac{\beta L}{4R^2} - \frac{L}{2\alpha\beta R^4} = 0, \quad (9.48)$$

and solve for R^2,

$$R^2 = \frac{\beta/4 \pm \sqrt{(\beta/4)^2 + 4\mu \frac{1}{2\alpha\beta}}}{2\mu}$$

$$= \frac{\beta^2 \alpha + \sqrt{\beta^4 \alpha^2 + 32\alpha\beta\mu}}{8\alpha\beta\mu}, \quad (9.49)$$

where we take the "plus" solution as giving physical results.

9.4 Monte Carlo Comparison

We apply Monte Carlo to the original quasi-2D model with Hamiltonian (9.3) to verify two hypotheses:

1. that the 3-D effects, namely the (9.27), predicted in the mean-field are correct
2. that these effects can be considered physical in the sense that the model's asymptotic assumptions of straightness is not violated.

Research on flux-lines in type-II superconductors has yielded a close correspondence between the behavior of vortex filaments in 3-space and paths of quantum bosons in (2+1)-D (2-space in imaginary time) [114, 135]. This work is not related to ours fundamentally because type-II superconductor flux-lines do not have the same boundary conditions. They use periodic boundaries in all directions with an interaction cut-off distance while we use no boundary conditions and no cut-off. Besides the boundaries, they also allow flux-lines to permute like bosons, switching the top end points, which we do not allow for our vortices. However, despite the boundary differences, the London free-energy functional for interacting flux-lines is closely related to our Hamiltonian (9.3), and so we can apply Path Integral Monte Carlo (PIMC) in the same way as it has been applied to flux-lines. (For a discussion of PIMC and how we apply it see Sect. 8.2.3.)

We simulated a collection of $N = 20$ vortices each with a piecewise linear representation with $M = 1,024$ segments and ran the system to equilibration, determined by the settling of the mean and variance of the total energy. We ran the system for 20 logarithmically spaced values of β between 0.001 and 1 plus two points, 10 and 100. We set $\alpha = 10^7$, $\mu = 2,000$, and $L = 10$. We calculate several arithmetic averages: the mean square vortex position,

$$R_{MC}^2 = (MN)^{-1} \sum_{i=1}^{N} \sum_{k=1}^{M} |\psi_i(k)|^2, \qquad (9.50)$$

where k is the segment index corresponding to discrete values of σ, and the mean square amplitude per segment,

$$a^2 = (MN)^{-1} \sum_{i=1}^{N} \sum_{k=1}^{M} |\psi_i(k) - \psi_i(k+1)|^2, \qquad (9.51)$$

where $\psi_i(M+1) = \psi_i(1)$.

Measures of (9.50) correspond well to (9.27) in Fig. 9.2 whereas (9.1) continues to decline when the others curve with decreasing β values, suggesting that the 3-D effects are not only real in the Monte Carlo but that the mean-field is a good approximation with these parameters.

In order to be considered straight enough, we need

$$a \ll \frac{L}{M} = \frac{10}{1024}. \qquad (9.52)$$

Straightness holds for all β values.

The Monte Carlo simulation begins with a random distribution of filament endpoints in a square of side 10, and there are two possible moves that the algorithm chooses at random:

1. Moves a filament's end-points, $\psi_i(1)$ and $\psi_i(M+1)$. The index i is chosen at random, and the filament i's end-points moved a uniform random distance.
2. Keeps end-points stationary and, following the bisection method of Ceperley (Fig. 8.2), grows a new internal configuration for a randomly chosen filament [30].

In each case, the energy of the new state, s', is calculated and retained with probability

$$A(s \to s') = \min\left\{1, \exp\left(-\beta[E_N^{s'}(M) - E_N^s(M)] - \mu[M_N^{s'}(M) - M_N^s(M)]\right)\right\}, \qquad (9.53)$$

where s is the previous state. This effectively samples states from the Gibbs probability distribution.

9.4 Monte Carlo Comparison

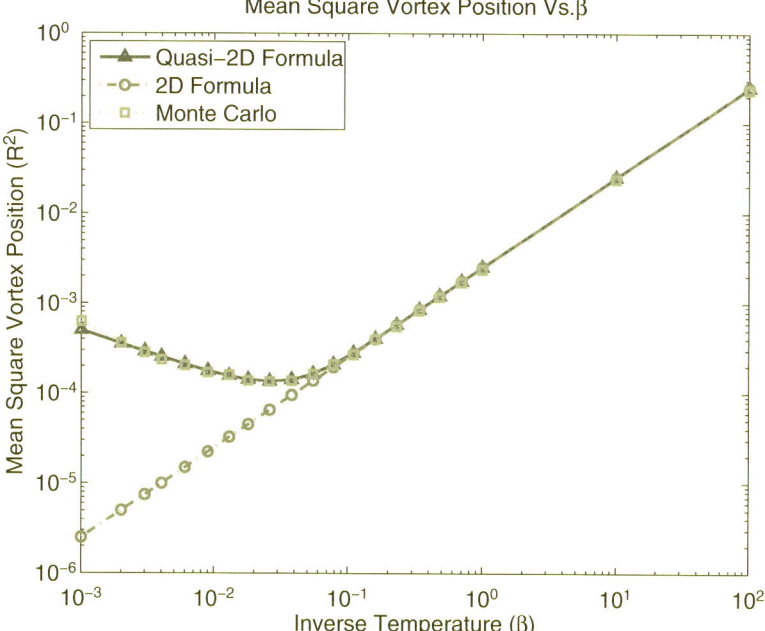

Fig. 9.2: The mean square vortex position, defined in (9.50), compared with (9.27) and (9.1) shows how 3-D effects come into play around $\beta = 0.16$. That the 2D formula continues to decrease while the Monte Carlo and the quasi-2D formula curve upwards with decreasing β suggests that the internal variations of the vortex lines have a significant effect on the probability distribution of vortices

Our stopping criteria is graphical in that we ensure that the cumulative arithmetic mean of the energy,

$$E_{\text{cum}}^k = k^{-1} \sum_{i=1}^k E_N(s_i) + \frac{\mu}{\beta} M_N(s_i), \qquad (9.54)$$

where s_i refers to the state resulting from the ith move and k is the current move index, settles to a constant. The energy is almost guaranteed to settle in the case of the Gibbs' measure because of the tendency for the system to select a particular energy state (mean energy) and remain close to that state. Typically, we run for 1 million moves (accepted plus rejected) or 50,000 sweeps for 20 vortices. Afterwards, we collect data from about 200,000 moves (1,000 sweeps) to generate statistical information.

9.5 Related Work

The closest study to this is the heton model study of [85] which applied a two layered vortex model to deep ocean convection. However, a two layered model is insufficient to show the effects we have shown. Briefly, we can mention a way to add the neglected planetary effect of the heton model to our filament model. In simple form, the rotation of the Earth can be taken as an additional vorticity term. Take each filament i to have a vorticity vector function, $\hat{\omega}_i(\tau)$, on some part of a rotating sphere. Rather than the axis of convection rotation being the z-axis, make the z-axis the axis of rotation of the sphere so that the sphere has vorticity $\Omega \hat{k}$, and give the filaments another axis of rotation, $\hat{t}$, somewhere on the sphere. The vorticity vector functions, $\hat{\omega}_i \tau$, are nearly parallel to $\hat{t}$ for all i. Call $\hat{\omega}_i(\tau)$ the *relative vorticity* and the *absolute vorticity* of each filament $\hat{q}_i(\tau) = \hat{\omega}_i(\tau) + \Omega \hat{k}$. This would effectively create a Coriolis effect in the filaments so that their behavior would be dependent on their location on the Earth's surface.

As mentioned in the previous section, simulations of flux lines in type-II superconductors using the PIMC method have been done, generating the Abrikosov lattice [114, 135]. However, the superconductor model has periodic boundary conditions in the xy-plane, is a different problem altogether, and is not applicable to trapped fluids. No Monte Carlo studies of the model of [78] have been done to date and dynamical simulations have been confined to a handful of vortices. Kevlahan [72] added a white noise term to the KMD Hamiltonian, (9.3), to study vortex reconnection in comparison with direct Navier–Stokes, but he confined his simulations to two vortices. Direct Navier–Stokes simulations of a large number of vortices are beyond our computational capacities.

Tsubota et al. [145] has done some excellent simulations of vortex tangles in He-4 with rotation, boundary walls, and *ad hoc* vortex reconnections to study disorder in rotating superfluid turbulence. Because vortex tangles are extremely curved, they applied the full Biot–Savart law to calculate the motion of the filaments in time. Their study did not include any sort of comparison to 2-D models because for most of the simulation vortices were far too tangled. The inclusion of rigid boundary walls, although correct for the study of He-4, also makes the results only tangentially applicable to the KMD system we use.

Our use of the spherical model is recent and has also been applied to the statistical mechanics of macroscopic fluid flows in order to obtain exact solutions for quasi-2D turbulence [86, 90].

Other related work on the statistical mechanics of turbulence in 3-D vortex lines can be found in [19, 51, 92].

9.6 Conclusion

We have developed an explicit mean-field formula for the most significant statistical moment for the quasi-2D model of nearly parallel vortex filaments and shown that in Monte Carlo simulations this formula agrees well while the related 2-D formula

9.6 Conclusion

fails at higher temperatures. We have also shown that our predictions do not violate the model's asymptotic assumptions for a range of inverse temperatures. Therefore, we conclude that these results are likely physical. We consider this strong evidence supporting our original hypothesis.

The implication toward deep ocean convection is that 3D effects become significant when vortex structures move close together and that this ultimately causes an expansion in the system size that one would not see in purely 2D structures. However, although the system as a whole expands, we cannot say whether this expansion is uniform or if there is a separation effect in which a small core surrounded by a halo of vortex structures emerges. Knowing that could have significant implications for the understanding of these structures.

Chapter 10
Conclusion

This book has two thrusts relating to vortex filaments in statistical equilibrium: (1) mathematical and computational approaches and (2) applications. The first thrust emphasizes the complementary nature of analysis and computation. It is wrong to assume that, because finite difference or element methods for solving the Navier–Stokes' equations or other PDEs are able to simulate vortex systems with high fidelity, that analytical approaches or Monte Carlo are obsolete. In high dimensional systems, numerical PDEs are not a reliable or tractable way of modeling systems. Furthermore, Monte Carlo, while valuable, cannot possibly replace analysis because parameter spaces of realistic systems are too large. Also, only analysis can provide rigorous proof of the existence of phenomena, particularly phase transitions which require a continuum, observed in simulation. Meanwhile, analysis alone runs into tractability problems that frequently require approximation, hence the need for complementary computational approaches. Indeed, in the preceding chapters analytical results have often been motivated by computational results.

Vortex filaments are an important area of research and likely to remain so because they appear in such a wide variety of applications from condensed matter physics to plasmas to ordinary fluids. Indeed, as William Thomson's work implies [143], vortex filaments are fundamental to matter because they are a unit of a conserved quantity, angular momentum, and thus reflect the rotational symmetry of space itself. As we have emphasized, treating vortex filaments as if they are self-contained quanta of angular momentum is a useful idea, even in non-quantum applications where vorticity is not quantized.

There are a few areas of research where vorticity and vortex filaments are starting to gain ground. The simulation of smoke and other non-solid phenomena which is useful both for understanding how smoke spreads and for computer animation [9, 150]. Vortex filaments primitives give artists a degree of control over their animations that other kinds of simulations do not, and there is a major interest in the computer graphics community in all things gaseous. The goal in smoke simulation is to obtain fast computation times with maximal resolution. A second goal is easy to obtain visually appealing simulations. High level primitives such as vortex filaments make this much easier. The simulation of filaments and rings interacting with smoke particles can create a variety of interesting visualizations.

Quantum vortices continue to be an important area of study. A recent study of quantum turbulence employed the full Biot–Savart law to study steady state counterflow [3], while turbulent Kelvin wave cascades on vortex filaments has been proposed as an explanation for turbulent decay in superfluids [14, 13]. As with smoke, vortex filaments aid visualizations of quantum turbulence [63]. Quantum turbulence (QT) is in many ways simpler than classical because of quantization of fluid rotation. Thus, QT offers a way to understand turbulence by seeing its "skeleton" in the form of intertwined vortex filaments. Because of this, vortex filament simulations take on a new importance because of the potential to unlocking the single remaining mystery of classical physics: turbulent flow. As superfluids and dilute BECs offer ways of testing hypotheses about turbulence, future research in this direction may unlock the secret of this phenomenon.

Vortex filaments have been studied for over a century, dating back to Thomson and Helmholtz, and we expect research to continue for another century at least. They are essential features of nature, quanta of angular momentum and/or magnetic fields as basic to the structure of the universe as elementary particles. The purpose of this book has been to discuss and elucidate some of the wider and more well-known points while also providing some deeper insights into the nature of vortex filaments, particularly in equilibrium. There is, however, a great deal more to be discovered, both in the ever growing literature on vortex filaments and research yet to be carried out. Future discoveries may find that vorticity lurks in deeper regions of physics such as within the hearts of elementary particles or the fabric of spacetime itself. If so, these observations will join a vast and growing understanding of these fundamental particles of rotation.

References

1. J.R. Abo-Shaeer, C. Raman, J.M. Vogels, W. Ketterle, Observation of vortex lattices in bose-einstein condensates. Science **292**, 476 (2001)
2. A.A. Abrikosov, On the magnetic properties of superconductors of the second group. Sov. Phys. JETP **5**(6), 1442–1452 (1957)
3. H. Adachi, S. Fujiyama, M. Tsubota, Steady-state counterflow quantum turbulence: simulation of vortex filaments using the full biot-savart law. Phys. Rev. B **81**(10), 104511 (2010)
4. A. Aftalion, I. Danaila, Giant vortices in combined harmonic and quartic traps. Phys. Rev. A **69**(033608) (2004)
5. O. Alexandrova, Solar wind vs magnetosheath turbulence and alfvn vortices. Nonlinear Process. Geophys. **15**(1), 95–108 (2008)
6. T.D. Andersen, C.C. Lim, Negative specific heat in generalized quasi-2d vorticity. Phys. Rev. Lett. **99**, 165001 (2007)
7. T.D. Andersen, C.C. Lim, Explicit formulae for nearly parallel vortex filaments. Geo. Astro. Fluid Dyn. **102**, 265–280 (2008)
8. T.D. Andersen, C.C. Lim, A length scale formula for confined quasi-2d plasmas. J. Plasma Phys. **75**, 437–454 (2009)
9. A. Angelidis, F. Neyret, Simulation of smoke based on vortex filamet primitives, in *Proceedings of the 2005 ACM SIGGRAPH/Eurographics symposium on Computer animation*, ACM, pp. 87–96 (2005)
10. J.F. Annett, *Superconductivity, Superfluids, and Condensates* (Oxford University Press, London, 2004)
11. I.S. Aranson, L. Kramer, The world of the complex ginzburg–landau equation. Rev. Mod. Phys. **74**, 99–143 (2002)
12. S.M. Assad, C.C. Lim, Statistical equilibrium distributions of baroclinic vortices in a rotating two-layer model at low froude numbers. Geophys. Astron. Fluid Dyn. **100**, 1–22 (2006)
13. A.W. Baggaley, C.F. Barenghi, Spectrum of turbulent kelvin-waves cascade in superfluid helium. Phys. Rev. B **83**(13), 134509 (2011)
14. A.W. Baggaley, C.F. Barenghi, Turbulent cascade of kelvin waves on vortex filaments, in *Journal of Physics: Conference Series*, vol 318 (IOP Publishing, 2011), p. 062001
15. J.L.F. Barbón, Black holes, information and holography. J. Phys. **171**, 012009 (2009)
16. C.F. Barenghi, R.J. Donnelly, W.F. Vinen (eds.), *Quantized Vortex Dynamics and Superfluid Turbulence* (Springer, Berlin, 2001)
17. C.F. Barenghi, D.C. Samuels, R.L. Ricca, Complexity measures of tangled vortex filaments, in *Tubes, Sheets and Singularities in Fluid Dynamics* (Springer, New York, 2002), pp. 69–74
18. S.M. Belotserkovsky, I.K. Lifanov, *Method of Discrete Vortices*, chap. 9 (CRC, Boca Raton, FL, 1993)
19. V. Berdichevsky, Statistical mechanics of vortex lines. Phys. Rev. E **57**, 2885 (1998)

20. T.H. Berlin, M. Kac, The spherical model of a ferromagnet. Phys. Rev. **86**(6), 821 (1952)
21. G.P. Bewley, M.S. Paoletti, K.R. Sreenivasan, D.P. Lathrop, Characterization of reconnecting vortices in superfluid helium. Proc. Natl. Acad. Sci. **105**(37), 13707–13710 (2008)
22. G. Blatter, M.V. Feigel'man, V.B. Geshkenbein, A.I. Larkin, V.M. Vinokur, Vortices in high-temperature superconductors. Rev. Mod. Phys. **66**(4), 1125–1388 (1994)
23. O.N. Boratav, R.B. Pelz, N.J. Zabusky, Reconnection in orthogonally interacting vortex tubes: direct numerical simulations and quantifications. Phys. Fluids A **4**(3), 581–605 (1992)
24. F. Bouchet, Simpler variational problems for statistical equilibria of the 2d euler equation and other systems with long range interactions. Phys. D **237**, 1976–1981 (2008)
25. F. Bouchet, A. Venaille, Statistical mechanics of two-dimensional and geophysical flows. Phys. Rep. **515**, 227–295 (2012)
26. L.S. Brown, *Quantum Field Theory* (Cambridge University Press, Cambridge, 1992)
27. D.J.E. Callaway, A. Rahman, Microcanonical ensemble formulation of lattice gauge theory. Phys. Rev. Lett. **49**(9), 613–616 (1982)
28. D.J.E. Callaway, A. Rahman, Lattice gauge theory in the microcanonical ensemble. Phys. Rev. D **28**(6), 1506–1514 (1983)
29. A.J. Callegari, L. Ting, Motion of a curved vortex filament with decaying vortical core and axial velocity. SIAM J. Appl. Math. **35**(1), 148–175 (1978)
30. D.M. Ceperley, Path integrals in the theory of condensed helium. Rev. Mod. Phys. **67**, 279 (1995)
31. P. Chatelain, A. Curioni, M. Bergdorf, D. Rossinelli, W. Andreoni, P. Koumoutsakos, Billion vortex particle direct numerical simulations of aircraft wakes. Comput. Methods Appl. Mech. Eng. **197**(13), 1296–1304 (2008)
32. P.H. Chavanis, Dynamical and thermodynamical stability of two-dimensional flows: variational principles and relaxation equations. Eur. Phys. J. B **70**, 73 (2009)
33. A.J. Chorin, Numerical study of slightly viscous flows. J. Fluid Mech. **57**, 785–796 (1973)
34. A.J. Chorin, *Vorticity and Turbulence* (Springer, New York, 1994)
35. A.J. Chorin, J. Akao, Vortex equilibria in turbulence theory and quantum analogues. Phys. D **51**, 403 (1991)
36. I.P. Christiansen, Numerical simulation of hydrodynamics by the method of point vortices. J. Comput. Phys. **13**(3), 363–379 (1973)
37. CISL Cisl fy2009 annual report. Technical report, NCAR (2009)
38. R. Clausius, Xvi. on a mechanical theorem applicable to heat. Phil. Mag. Ser. 4 **40**(265), 122–127 (1870)
39. M. Creutz, Microcanonical monte carlo simulation. Phys. Rev. Lett. **50**, 1411–1414 (1983)
40. I. Danaila, Three-dimensional vortex structure of a fast rotating bose-einstein condensate with harmonic-plus-quartic confinement. Phys. Rev. A **72**, 013605 (2005)
41. I. Danaila, F. Hecht, A finite element method with mesh adaptivity for computing vortex states in fast-rotating boseeinstein condensates. J. Comput. Phys. **229**, 6946–6960 (2010)
42. I.V. Despirak, A.A. Lubchich, V. Yu. Trakhtengerts, Excitation of alfvn vortices in the ionosphere by the magnetospheric convection. Radiophys. Quantum Electron. **51**(5), 339–351 (2008)
43. M.T. Di Battista, A. Majda, Equilibrium statistical predictions for baroclinic vortices: the role of angular momentum. Theor. Comput. Fluid Dyn. **14**:293 (2001)
44. X. Ding, C.C. Lim, Phase transitions of the energy-relative enstrophy theory for a coupled barotropic fluid—rotating sphere system. Phys. A **374**, 152–164 (2007)
45. R. Eckhardt, Stan ulam, john von neumann, and the monte carlo method. Los Alamos Sci. **15**, 131–137 (1987)
46. S.F. Edwards, J.B. Taylor, Negative temperature states for a two-dimensional plasmas and vortex fluids. Proc. R. Soc. Lond. A **336**, 257–271 (1974)
47. R.P. Feynman, *Progress in Low Temperature Physics* (North-Holland, Amsterdam, 1955)
48. R.P. Feynman, J.W. Wheeler, Space-time approach to non-relativistic quantum mechanics. Rev. Mod. Phys. **20**, 367 (1948)

References

49. M.P.A. Fisher, Vortex-glass superconductivity: a possible new phase in bulk high-tc oxides. Phys. Rev. Lett. **62**, 1415–1418 (1989)
50. R. Fitzpatrick, The physics of plasmas, Lecture notes (2010)
51. F. Flandoli, M. Gubinelli, The gibbs ensemble of a vortex filament. Probab. Theory Rel. Fields **122**(2), 317 (2002)
52. P.J. Forrester, S.O. Warnaar, The importance of the selberg integral. arxiv, (0710.3981) (2007)
53. J.W. Gibbs, *Elementary Principles in Statistical Mechanics* (Charles Scribner Sons, New York, 1902)
54. S.B. Giddings, The black hole information paradox, in *Particles, Strings and Cosmology*, Johns Hopkins Workshop on Current Problems in Particle Theory 19 and the PASCOS Interdisciplinary Symposium 5 (1995)
55. J. Ginibre, Statistical ensembles of complex, quaternion, and real matrices. J. Math. Phys. **6**(3), 440–449 (1965)
56. V.L. Ginzburg, Nobel Lecture: On superconductivity and superfluidity (what I have and have not managed to do) as well as on the "physical minimum" at the beginning of the XXI century. Rev. Mod. Phys. **76**(3), 981 (2004)
57. V.L. Ginzburg, L.D. Landau, in *Collected Papers of L. D. Landau* (Pergamon Press, Oxford, 1965), p. 546
58. G. Glatzmaier, R. Coe (2007) 8.09—Magnetic polarity reversals in the core, in *Gerald Schubert, editor, Treatise on Geophysics* (Elsevier, Amsterdam, 2007), pp. 283–297
59. G. Glatzmaier, P.H Roberts, A three-dimensional self-consistent computer simulation of a geomagnetic field reversal. Nature **377**, 203–209 (1995)
60. P.E. Goa, H. Hauglin, M. Baziljevich, E. Il'yashenko, P.L. Gammel, T.H. Johansen, Real-time magneto-optical imaging of vortices in superconducting nbse2. Superconductor Sci. Technol. **14**(9), 729 (2001)
61. A.V. Gordeev, A.S. Kingsep, L.I. Rudakov, Electron magnetohydrodynamics. Phys. Rep. **243**, 215 (1994)
62. E.P. Gross, Structure of a quantized vortex in boson systems. Nuevo Cimento **20**, 454 (1961)
63. R. Hänninen, A.W. Baggaley, Vortex filament method as a tool for computational visualization of quantum turbulence. Proc. Natl. Acad. Sci. **111**(Supplement 1), 4667–4674 (2014)
64. J.W. Hartman, P.B. Weichman, The spherical model for a quantum spin glass. Phys. Rev. Lett. **74**(23), 4584 (1995)
65. H. Hasimoto, A soliton on a vortex filament. J. Fluid Mech. **51**, 472 (1972)
66. W.K. Hastings, Monte Carlo sampling methods using Markov chains and their applications. Biometrika **57**, 97–109 (1970)
67. C. Herbert, Phase transitions and marginal ensemble equivalence for freely evolving flows on a rotating sphere. Phys. Rev. E **85**, 056304 (2012)
68. C. Herbert, Statistical mechanics of quasi-geostrophic flows on a rotating sphere. J. Stat. Mech. **5**, P05023 (2012)
69. G. Horwitz, Steepest descent path for the microcanonical ensemble- resolution of an ambiguity. Commun. Math. Phys. **89**, 117–129 (1983)
70. G.R. Joyce, D. Montgomery, Negative temperature states for a two-dimensional guiding center plasma. J. Plasma Phys. **10**, 107 (1973)
71. J. Keller, W. Pressman, Equation of state and phase transition of the spherical lattice gas. Phys. Rev. **120**, 22–32 (1960)
72. N.K.-R. Kevlahan, Stochastic differential equation models of vortex merging and reconnection. Phys. Fluids **17**, 065107 (2005)
73. M.K.-H. Kiessling, T. Neukirch, Negative specific heat of a magnetically self-confined plasma torus. PNAS **100**, 1510–1514 (2003)
74. R. Kinney, T. Tajima, N. Petviashvili, Discrete vortex representation of magnetohydrodynamics. Phys. Rev. Lett. **71**, 1712–1715 (1993)
75. R. Kinney, T. Tajima, J.C. McWilliams, N. Petviashvili, Filamentary magnetohydrodynamic plasmas. Phys. Plasmas (1994-present) **1**(2), 260–280 (1994)

76. D. Kleckner, W.T.M. Irvine, Creation and dynamics of knotted vortices. Nat. Phys. **9**, 253–258 (2013)
77. R. Klein, A.J. Majda, Self-stretching of a perturbed vortex filament i: the asymptotic equation for deviations from a straight line. Phys. D **49**, 323 (1991)
78. R. Klein, A. Majda, K. Damodaran, Simplified equation for the interaction of nearly parallel vortex filaments. J. Fluid Mech. **288**, 201–48 (1995)
79. T. Klein, I. Joumard, S. Blanchard, J. Marcus, R. Cubitt, T. Giamarchi, P.L. Doussal, A bragg glass phase in the vortex lattice of a type ii superconductor. Nature **413**(6854), 404–406 (2001)
80. A.N. Kolmogorov, Local structure of turbulence in an incompressible fluid at very high reynolds number. Dokl. Akad. Nauk SSSR **30**, 299–302 (1941)
81. R.H. Kraichnan, Statistical dynamics of two-dimensional flows. J. Fluid Mech. **67**, 155–175 (1975)
82. H. Lamb, *Hydrodynamics* (Cambridge University Press, Cambridge, 1916)
83. C. Leith, Minimum enstrophy vortices. Phys. Fluids **27**, 1388–1395 (1984)
84. Y. Levin, R. Pakter, F.B. Rizzato, T.N. Teles, F.P.C. Benetti, Nonequilibrium statistical mechanics of systems with long-range interactions. Phys. Rep. **535**(1), 1–60 (2014)
85. C. Lim, A. Majda, Point vortex dynamics for a coupled surface interior qg and propagating heton clusters in models for ocean convection. Geophys. Astron. Fluid Dyn. **94**, 177–220 (2001)
86. C.C. Lim, Phase transitions to super-rotation in a coupled barotropic fluid rotating sphere system, in *Proc. IUTAM Symp.*, Steklov Inst., Moscow, August 2006. Plenary Talk in Proc. IUTAM Symp., (Springer, New York, 2007)
87. C.C. Lim, Phase transition to super-rotating atmospheres in a simple planetary model for a non-rotating massive planet - exact solution. Phys. Rev E **86**, 066304 (2012)
88. C.C. Lim, S.M. Assad, Self-containment radius for rotating planar flows, single-signed vortex gas and electron plasma. R & C Dyn. **10**, 240–254 (2005)
89. C.C. Lim, X. Ding, J. Nebus, *Vortex Dynamics, Statistical Mechanics, and Planetary Atmospheres* (World Scientific, Singapore, 2009)
90. C.C. Lim, J. Nebus, *Vorticity Statistical Mechanics and Monte-Carlo Simulations* (Springer, New York, 2006)
91. D. Lindley, Boltzmann's atom: the great debate that launched a revolution in physics. Simon and Schuster (2001)
92. P.-L. Lions, A.J. Majda, Equilibrium statistical theory for nearly parallel vortex filaments. Comm. Pure Appl. Math. **LIII**, 76–142 (2000)
93. T.S. Lundgren, Y.B. Pointin, Statistical mechanics of two-dimensional vortices. J. Stat. Phys. **17**, 323–355 (1977)
94. D. Lynden-Bell, R.M. Lynden-Bell, On the negative specific heat paradox. Mon. Not. R. Astron. Soc. **181**, 405–419 (1977)
95. D. Lynden-Bell, R. Wood, The gravo-thermal catastrophe in isothermal spheres and the onset of red-giant structure for stellar systems. Mon. Not. R. Astron. Soc. **138**, 495–525 (1968)
96. A. Majda, X. Wang, *Non-linear Dynamics and Statistical Theories for Basic Geophysical Flows* (Cambridge University Press, Cambridge, 2006)
97. A. Majda, X. Wang, *Nonlinear Dynamics and Statistical Theories for Basic Geophysical Flows* (Cambridge University Press, Cambridge, 2008)
98. J.B. Marston, Planetary atmospheres as non-equilibrium condensed matter. Ann. Rev. Cond. Matter Phys. **3**, 285 (2012)
99. J.B. Marston, W. Qi, Hyperviscosity and statistical equilibria of euler turbulence on the torus and the sphere (2013). preprint arXiv:1312.2553
100. R. Matsumoto, T. Tajima, K. Shibata, M. Kaisig, Three-dimensional magnetohydrodynamics of the emerging magnetic flux in the solar atmosphere. Astrophys. J. **414**, 357–371 (1993)
101. J.C. Maxwell, *Theory of Heat* (D. Appleton and Co., New York, 1872)
102. T. Maxworthy, S. Narimousa, Unsteady, turbulent convection into a homogeneous, rotating fluid with oceanographic applications. J. Phys. Oceanogr. **24**, 865 (1994)

References 135

103. D.A. McQuarrie, *Statistical Mechanics* (University Science Books, Sausalito, 2000)
104. M.L. Mehta, *Random Matrices*, 3rd edn. (Academic Press, Amsterdam, 2004)
105. N. Metropolis, A.W. Rosenbluth, M.N. Rosenbluth, A.H. Teller, E. Teller, Equations of state calculations by fast computing machines. J. Chem. Phys. **21**, 1087–1092 (1953)
106. J. Miller, Statistical mechanics of euler equations in two dimensions. Phys. Rev. Lett. **65**, 2137–2140 (1990)
107. J. Miller, P.B. Weichman, M.C. Cross, Statistical mechanics, Euler's equation, and Jupiter's red spot. Phys. Rev. A **45**, 2328–2361 (1992)
108. H.K. Moffatt, The degree of knottedness of tangled vortex lines. J. Fluid Mech. **35**(01), 117–129 (1969)
109. H.K. Moffatt, R.L. Ricca, Helicity and the calugareanu invariant. Proc. R. Soc. Lond. Ser. A **439**(1906), 411–429 (1992)
110. Y. Nakamura, H. Bailung, P.K. Shukla, Observation of ion-acoustic shocks in a dusty plasma. Phys. Rev. Lett. **83**, 1602–1605 (1999)
111. T. Nattermann, S. Scheidl, Vortex-glass phases in type-ii superconductors. Adv. Phys. **49**(5), 607–704 (2000)
112. J.C. Neu, Vortices in complex scalar fields. Phys. D **43**, 385–406 (1990)
113. P.K. Newton, *The N-vortex Problem*, vol. 145 of Appl. Math. Sci. (Springer, New York, 2001)
114. H. Nordborg, G. Blatter, Numerical study of vortex matter using the bose model: first-order melting and entanglement. Phys. Rev. B **58**(21), 14556 (1998)
115. L. Ofman, B.J. Thompson, Sdo/aia observation of kelvinhelmholtz instability in the solar corona. Astrophys. J. Lett. **734**(1), L11 (2011)
116. A. Ogawa, *Vortex Flow* (CRC, Boca Raton, 1993)
117. L. Onsager, Statistical hydrodynamics. Nuovo Cimento Suppl. **6**, 279–287 (1949)
118. L. Onsager, P.C. Hammer, H. Holden, S.K. Ratkje, *The Collected Works of Lars Onsager: with Commentary* (World Scientific, Singapore, 1996), p. 744
119. W. Pauli, *Statistical Mechanics* (MIT Press, Cambridge, 1973)
120. L.M. Pismen, *Vortices in Nonlinear Fields* (Clarendon Press, Oxford, 1999)
121. L.P. Pitaevskii, Vortex lines in imperfect bose gas. Sov. Phys. JETP **13**, 451 (1961)
122. J. Preskill, Do black holes destroy information? in *International Symposium on Black Holes, Membranes, Wormholes, and Superstrings* (1992)
123. E.M. Purcell, R.V. Pound, A nuclear spin system at negative temperature. Phys. Rev. **81**, 279–280 (1951)
124. S. Raasch, D. Etling, Modeling deep ocean convection: large eddy simulation in comparison with laboratory experiments. J. Phys. Oceanogr. **28**, 1786 (1998)
125. N. Ramsey, Thermodynamics and statistical mechanics at negative absolute temperatures. Phys. Rev. **103**(1), 20–28 (1956)
126. R.L. Ricca, Physical interpretation of certain invariants for vortex filament motion under lia. Phys. Fluids A **4**(5), 938–944 (1992)
127. R.L. Ricca, Structural complexity and dynamical systems, in *Lectures on Topological Fluid Mechanics* (Springer, New York, 2009), pp. 167–186
128. R.L. Ricca, M.A. Berger, Topological ideas and fluid mechanics. Phys. Today **49**(12), 28–34 (2008)
129. R.S. Richardson, The nature of solar hydrogen vortices. Astrophys. J. **93**, 24 (1941)
130. R. Robert, J. Sommeria, Statistical equilibrium states for two-dimensional flows. J. Fluid Mech. **229**(29), 1–3 (1991)
131. P.G. Saffman, *Vortex Dynamics* (Cambridge University Press, Cambridge, 1995)
132. T. Sakajo, Transition of global dynamics of a polygonal vortex ring on a sphere with pole vortices. Phys. D **196**(3–4), 243–264 (2004)
133. E. Schrödinger (ed.), *Statistical Thermodynamics* (Dublin Institute for Advanced Studies, Dublin, 1944)
134. E. Schrödinger, *Statistical Thermodynamics* (Cambridge University Press, Cambridge, 1952)
135. P. Sen, N. Trivedi, D.M. Ceperley, Simulation of flux lines with columnar pins: bose glass and entangled liquids. Phys. Rev. Lett. **86**(18), 4092 (2001)

136. Z.S. She, E. Jackson, S.A. Orszag, Statistical aspects of vortex dynamics in turbulence, In *New Perspectives in Turbulence* (Springer, New York, 1991), pp. 315–328
137. P.K. Shukla, Dust ion-acoustic shocks and holes. Phys. Plasmas (1994-present) **7**(3), 1044–1046 (2000)
138. P.K. Shukla, A survey of dusty plasma physics. Phys. Plasmas **8**(5), 1791–1803 (2001)
139. L. Sirovich, *Techniques of Asymptotic Analysis*. Applied Mathematical Sciences (Springer, New York, 1971)
140. M.A. Sokolovskiy, J. Verron, *Dynamics of Vortex Structures in a Stratified Rotating Fluid*, vol. 47 of Atmospheric and Oceanographic Sciences (Springer, New York/Heidelberg, 2014)
141. H.E. Stanley, Spherical model as the limit of infinite spin dimensionality. Phys. Rev. **176**, 718 (1968)
142. W. Thirring, Systems with negative specific heat. Z. Phys. **235**, 339 (1970)
143. W. Thomson, Ii. on vortex atoms. Lond. Edinb. Dublin Philos. Mag. J. Sci. **34**(227), 15–24 (1867)
144. L. Ting, R. Klein, *Viscous Vortical Flows*, vol. 374 of Lecture Notes in Physics (Springer, Berlin, 1991)
145. M. Tsubota, T. Araki, C.F. Barenghi, Rotating superfluid turbulence. Phys. Rev. Lett. **90**, 205301 (2003)
146. B. Turkington, R. Ellis, An introduction to the thermodynamic and macrostate levels of nonequivalent ensembles. Phys. A **340**, 138–146 (2004)
147. L. Uby, M.B. Isichenko, V.V. Yankov, Vortex filament dynamics in plasmas and superconductors. Phys. Rev. E **52**, 932–939 (1995)
148. A. van Otterlo, R.T. Scalettar, G.T. Zimányi, Phase diagram of disordered vortices from london langevin simulations. Phys. Rev. Lett. **81**(7), 1497 (1998)
149. S. von Goeler, W. Stodiek, and N. Sauthoff, Studies of internal disruptions and $m=1$ oscillations in tokamak discharges with soft X-ray techniques. Phys. Rev. Lett. **33**, 1201–1203 (1974)
150. S. Weißmann, U, Pinkall, Filament-based smoke with vortex shedding and variational reconnection. ACM Trans. Graph. (TOG) **29**(4), 115 (2010)
151. J. Wesson, The science of jet. Technical Report JET-R(99)13, JET (1999)
152. A. Zee, *Quantum Field Theory in a Nutshell* (Princeton University Press, Princeton, 2003)

Index

Symbols
2D point vortex gas, 44

A
Abrikosov
    lattice, 19
    Nobel prize, 19
Abrikosov lattice, 19, 126
    decoration experiment, 19
Alfven's theorem, 68
asymptotic matching, 57, 58, 62
atmospheric super-rotation, 10
    phase transitions, 12

B
bathtub vortex, 15
    Coriolis force, 15
Biot-Savart Law, 2
Biot-Savart law, 15, 45, 59
Boltzmann's constant, 33
Bose–Einstein condensate
    liquid Helium-4, 17
Bose-Einstein condensate, 17
    dilute, 17
    quantized vortices in, 18
    two-fluid model, 17

C
calculus of variations, 32
    pitfalls, 36
Callegari and Ting, 59
canonical ensemble, 27, 34
    equivalence with microcanonical, 40
Ceperley, David, 104
circulation, 10

complex order parameter, 75
configuration space, 37

D
deep ocean convection, 15, 111
Demon algorithm, 107
density function, 32
dimensionality, the curse of, 102
discrete vortices method, 14
dust ion-acoustic holes, 23
dust ion-acoustic shocks, 23

E
entropy
    Gibbs definition, 33
equipartition theorem, 36
ergodicity, 103
Euler equations, 1

F
flux freezing equation, 21
frozen field equation, *see* flux freezing equation

G
Gibbs entropy, 33
Ginzburg-Landau theory, 75
grand canonical ensemble, 27
Great Dark Spot of Neptune, 13
Great Red Spot of Jupiter, 12
Great White Spot of Saturn, 12
Gross-Pitaevskii equation, 72
guiding center model, 3, 51
    state equations, 52
gyro-radius, *see* Larmor radius

H
Hamiltonian flow algorithms, 108
Heisenberg model, 27
Helmholtz free energy, 34
Helmholtz's laws of vortex motion, 2, 59, 61
Huygens probe, 12
Hydrodynamics, 9

I
integral approach, 37
Ising model, 27

K
Kelvin-Helmoltz instability, 23
Kolmogorov energy cascade, 50

L
Lagrange multipliers, 33
Lamb, Horace, 43
Langmuir, Irving, 21
Larmor radius, 82
laws of vortex motion, *see* Helmholtz's laws of vortex motion
LIA, *see* local induction approximation
Liouville theorem, 26
local induction approximation, 66, 112
London vortices, 76

M
macrostate, 31
magnetic nuclear fusion, 51
Magnetic pole reversal, 21
Magnetohydrodynamic plasma model, 21
Mariner 10, 11
matched expansion, *see* asymptotic matching
Maxwell's Demon, 106
MCMC, *see* Markov Chain Monte Carlo
Meissner effect, 19
meta stability, 89
Metropolis algorithm, 102
microcanonical ensemble, 28, 34
  equivalence with canonical, 40
microcanonical partition function, 39
microstate, 31
Miller-Robert model, 54
minimalist approach, 27

N
Navier-Stokes equations, 1, 9
  vorticity form, 10
nearly parallel vortex filaments, 65
  Hamiltonian, 66
  mean field theory, 116
negative specific heat, 7, 89

negative temperature, 47
negative temperature states, 3
negative temperatures, 47
  Onsager's approach, 50
  properties of, 48
no-slip, no-penetration boundary conditions, 13
non-extensive limit, 34
nuclear fusion, 3
nuclear spin systems, 47

O
Onsager gas, 44
Onsager, Lars, 3

P
partition function, 37
path integral Monte Carlo, 103, 126
  bisection move, 124
perfect conductivity equation, *see* flux freezing equation
phase volume, 26
PIMC, *see* path integral Monte Carlo
Pioneer Venus Orbiter, 11
plasmas, 20
Polar vortices, 13
pole reversal, 21
Potts model, 27

Q
quantized vortices, 16

R
Robert-Miller model, 54

S
saddle point method, *see* steepest-descent method
solar vortices, 23
specific heat
  positiveness in canonical ensemble, 38
spherical model, 87
statistical ensemble, 37
  canonical, 37
  definition of, 27
  microcanonical, 38
statistical mechanics
  active areas of research, 26
  definition of, 25
  equilibrium, 25
  justification for, 4
  two branches of, 25
steepest-descent method, 87, 119, 120

Index

Superconductor
 stoichiometric type-II, 18
Superconductors, 18
superconductors, 123
superfluid, 17
surfer's wave, *see* Kelvin-Helmholtz instability

T
temperature, 47
thermodynamic limit, 34
Titan, 12
trefoil vortex, 68
turbulence as an equilibrium phenomenon, 51
two-dimensional vortex gas
 equations of motion, 46
 guiding center model, 51
 Hamiltonian system, 46

U
ultraviolet catastrophe, 34

V
van der Waals model, 26
variational formula, 34
Venus Express, 11
vortex filaments, 1
 angular momentum, 37
 as holes in fluid, 37
 curvilinear form, 59
 in planetary atmospheres, 10
 negative specific heat in statistical
  equilibrium, 89
 radial pressure gradient, 58
 reconnection, 67
 two region model, 58
vortex glass, 78
vortex liquid, 19
vortex pinning, 19
vortices
 as particles, 5
 energy dissipation of, 5
 equilibrium time scale, 5
 filaments, *see* vortex filaments
 laws of motion, *see* Helmholtz's laws of
  vortex motion
 merger, 68
 quantized, 5, 16
 two-dimensional gas
  fixed energy radius, 7
  limitations of, 6
  negative specific heat, 7
Voyager 1, 12

W
wake vortices, 13

X
XY model, 27

MIX
Papier aus verantwortungsvollen Quellen
Paper from responsible sources
FSC® C105338

If you have any concerns about our products,
you can contact us on
ProductSafety@springernature.com

In case Publisher is established outside the EU,
the EU authorized representative is:
**Springer Nature Customer Service Center GmbH
Europaplatz 3, 69115 Heidelberg, Germany**

Printed by Libri Plureos GmbH
in Hamburg, Germany